Coast-to-Coast Auto Races of the Early 1900s:

Three Contests That Changed the World

Other SAE historical books

Carriages Without Horses: J. Frank Duryea and the Birth of the
American Automobile Industry
By Richard P. Scharchburg
(Order No. R-127)

The Franklin Automobile Company: The History of the Innovative
Firm, Its Founders, The Vehicles It Produced (1902-1934) and The
People Who Built Them
By Sinclair Powell
(Order No. R-208)

The Golden Age of the American Racing Car
By Griffith Borgeson
(Order No. R-196)

For more information or to order this book, contact SAE at 400 Commonwealth Drive,
Warrendale, PA 15096-0001; (724)776-4970; fax: (724)776-0790;
e-mail: publications@sae.org; web site: www.sae.org/BOOKSTORE.

Coast-to-Coast Auto Races of the Early 1900s:

Three Contests That Changed the World

Curt McConnell

SAE INTERNATIONAL ®
Society of Automotive Engineers, Inc.
Warrendale, Pa.

Library of Congress Cataloging-in-Publication Data

McConnell, Curt, 1959-
    Coast-to-Coast auto races of the early 1900s : three contests that changed the world / Curt McConnell.
        p. cm.
    Includes bibliographical references and index.
    ISBN 0-7680-0604-X
    1. Automobile racing--United States--History. 2. New York to Paris Race, 1908. I. Title.

GV1033.M39 2000
796.72'0973'09041--dc21   00-040054

Copyright © 2000   Society of Automotive Engineers, Inc.
                        400 Commonwealth Drive
                        Warrendale, PA 15096-0001 U.S.A.
                        Phone: (724)776-4841
                        Fax: (724)776-5760
                        E-mail: publications@sae.org
                        http://www.sae.org

ISBN 0-7680-0604-X

SAE Order No. R-275

Dedication

Dedicated to the memory of the adventuresome men who, despite overwhelming obstacles, forced their automobiles across the continent, competing not only against one another but against the very forces of nature.

> "The contestants … must be men who are not easily overcome by difficulties, who are strong mentally and physically; able to use an axe to good purpose; who can wade through water up to their armpits, if necessary, in seeking to pull a machine out of a 'bad place,' who can withstand cold and possibly hunger without taking it too much to heart … and who, in conclusion, can adapt themselves to circumstances. It will, therefore, be seen that as much, if not a great deal more, depends on the 'men behind the machine' than on the machine itself, for no matter how good an automobile entered for the race may be, unless there is the right kind of human intelligence, will, and stamina to guide it, success cannot be hoped for."

—The *New York Times*, outlining the traits necessary to win the 1908 New York–Paris race[1]

---

[1] "Making Ready to Race from New York to Paris," *New York Times*, Dec. 29, 1907, V, p. 1:2.

Contents

Photo Credits

Acknowledgments

I drove all over the country to write the story of three transcontinental races from the first decade of the twentieth century, seeing—albeit in much greater comfort—many of the same sights that the drivers themselves viewed. Back home again, attempting to flesh out the story, I wrote hundreds of letters seeking photographs, biographical information and answers to various questions, large and small. Reference librarians, as a group, fielded most of these questions; their replies helped me enormously in my quest to piece together the events of long ago. I have not space enough, unfortunately, to thank each reference librarian, curator or archivist by name.

In other instances, I must name names. For their editing assistance, I'd like to thank my father, Campbell R. McConnell, and my sisters, Lauren McConnell Davis and Beth McConnell. For general research assistance, or for help answering specific questions for more than one chapter, I gratefully acknowledge the contributions of Ralph Dunwoodie, Beverly Rae Kimes and David Smith, as well as:

- Kim Miller of the Antique Automobile Club of America's Library & Research Center, Hershey, Pennsylvania.
- Peggy Dusman, Dan Kirchner and Karen Prymak of the American Automobile Manufacturers Association, Detroit.
- Louis G. Helverson, Stuart McDougall and Bob Rubenstein of the Automobile Reference Collection, Free Library of Philadelphia.
- Jim Lesniak and Sandy Saunders of the National Automobile Museum, Reno, Nevada.
- Serena Gomez and Mark Patrick of the National Automotive History Collection, Detroit Public Library, and Tom Sherry, the collection's contract photographer.
- Roger White of the Transportation Division, Smithsonian Institution, Washington, D.C.
- The staffs of Bennett Martin Public Library, Lincoln, Nebraska, and Love Library, University of Nebraska-Lincoln.

Special thanks are due Aiyana and Mariah McConnell-Beepath, Ben and Holly Davis, Audrey and Stephen Broll, and my mother, Marilyn Knight McConnell. Among those who made vital contributions to specific chapters are these helpful persons:

Chapter 1, 1905 Olds race: Helen J. Earley and James R. Walkinshaw, Oldsmobile History Center, Lansing, Michigan; Jack Down, R.E. Olds Transportation Museum, Lansing, Michigan; and Edd Whitaker.

Chapter 2, 1908 New York–Paris race: Ralph Dunwoodie, John McCulloh, and Amanda Stanton-Geddes.

Chapter 3, 1909 New York–Seattle race: Richard Comeau; Joanne B. Harriman; Cliff Jenkins; Florence Pettengill; Hazel Pettengill; and Nick Scalera of the Henry Ford Museum & Greenfield Village, Dearborn, Michigan.

Introduction

Some 300,000 people gather in Speedway, Indiana, each Memorial Day weekend to watch 33 cars turn 200 laps around a $2^1/_2$-mile oval at speeds exceeding 200 mph. The Indianapolis 500, promoted as "the greatest spectacle in racing," lasts about three hours. If you watch the race on television, this includes well over 30 minutes of commercials.

What is at stake? Millions of dollars in sponsorships and prize money. What does it take to win? An experienced driver, a $350,000 race car, an annual budget of $4 million or more, countless hours of preparation, and a little bit of luck.

Is this real racing? By modern standards, it is. By the standards of the early 1900s, it is not, because the driver of the present-day Formula 1 auto is less important to the outcome of a race than is the technology that makes possible speeds of 225 mph. Formula 1 cars also bear no practical resemblance to street machines. Unlike its early-day counterpart, modern track racing has become predictable, almost civilized.

Years ago, the first man to fly solo across the Atlantic accurately forecast this sad state of affairs. "As time passes," Charles Lindbergh lamented, "the perfection of machinery tends to insulate man from contact with the elements in which he lives."[1]

In the first decade of the twentieth century, America saw the running of three grandiose automobile races—transcontinental races, held in 1905,1908 and 1909. During these contests, men had abundant contact with the elements: ice and snow, rain, hail, wind, heat, mud and dust. Races lasted not hours but weeks, pitted mere man against untamed nature, produced national heroes and galvanized the country like no modern race around an asphalt oval ever could.

In the process, these races sold America on the utility of the automobile, demonstrated beyond a doubt the need for good roads, and sparked a demand for everything from better autos to better road signs and maps. Furthermore, these races influenced not only potential automobile buyers but also officials with the power to shape the country's future, including the men who helped develop the Lincoln Highway, America's first transcontinental road for automobiles.

Yet writers of automotive history have largely ignored the coast-to-coast race of two curved-dash Oldsmobiles in 1905; disregarded the contestants who defied mud, mountains and "generalship" tactics—otherwise known as "dirty tricks"—in the 1909 New York–Seattle race; and (although to a lesser extent) overlooked the lessons of the wintertime crossing of America by the 1908 New York–Paris racers.

---

[1]Charles A. Lindbergh, Foreword to Anne Morrow Lindbergh, *Listen! the Wind* (New York: Harcourt, Brace and Co.,1938), pp. viii–ix.

Historians, instead, have tended to look at how track racing, hill climbs, economy contests, reliability runs, and other closely regulated events stimulated the public interest in the automobile. Why? These controlled contests are relatively easy to research through automobile journals alone. By contrast, writing about a coast-to-coast race requires a time-consuming search of daily and weekly newspapers along the route, both to learn what happened and to gauge public reaction to it.

Thus, the relative handful of automotive historians who have even considered transcontinental races have done so in a cursory fashion. Predictably, they have often erred in reporting the most basic details. In *The American Automobile: A Centenary, 1893–1993*, for instance, Nick Georgano names the wrong contestant as the winner of the 1909 New York–Seattle race.[2]

Although historians have been generally reluctant to explore the notion that coast-to-coast races helped sell Americans on the automobile, several leading authors have acknowledged the truth of the argument. In his book, *The Car Culture*, James J. Flink asserts that long-distance reliability runs did more than other stunts to fire the public's imagination. (Transcontinental races, of course, blended long-distance reliability with speed, endurance and daring.) Similarly, in *The American Automobile: A Brief History*, John B. Rae contends that "demonstrations of ability to cover long distances" were more important than track racing in winning wide acceptance for the automobile in America.[3]

Splashed across the pages of hundreds of newspapers, the daily accounts of these coast-to-coast races read like adventure serials. And why not? The racers faced conditions unknown to the average American newspaper reader. The men and their cars routinely stuck in the mud, fell through bridges, careened down mountainsides, smashed through tangles of sagebrush and stalled in the midst of rushing rivers.

The rigors of the road sickened all four contestants in the 1905 Oldsmobile race. Near Granger, Wyoming, one of the 1908 New York–Paris race cars slid into a muddy stream, where for 30 hours it resisted all rescue attempts. Freeing it required the help of 150 railroad laborers armed with chains and ropes. Another Paris-bound race car sank so deeply in a muddy road that the vehicle actually overturned. As the demoralized crewmen attempted to excavate their machine, the mud froze. A freight train hit and crippled one of the 1909 Seattle-bound transcontinental race cars. Its crew, for lack of an automobile bridge, was attempting to cross the Missouri River on a railroad trestle.

Racing across the continent was a young man's sport. The average age of the New York–Paris racers, for instance, was 27. Yet their experiences en route aged them perceptibly. "No more

---

[2]Nick Georgano, *The American Automobile: A Centenary,1893–1993* (New York: Smithmark, 1992), p. 37.

[3]See James J. Flink, *The Car Culture* (Cambridge, Mass.: MIT Press, 1975), p. 21; and John B. Rae, *The American Automobile: A Brief History* (Chicago: University of Chicago Press,1965), p. 30.

ocean to ocean races for me," vowed one heartbroken New York–Seattle racer. "We couldn't have worked harder if we carried the car across the country on our backs," another racer declared. But these were afterthoughts. Luckily for us, a handful of automobile racers— young, exuberant, determined and resourceful—did set off from New York City in three separate races to blaze a trail west.

And now, turn the page and flip back through the years to an age when speeds of 225 mph were a wild-eyed fantasy—when 225 miles per week was often all a driver could hope for.

CHAPTER 1

Coast-to-Coast Racing Stunt Expands Auto's Influence

We half waded and half swam to shore, leaving the car with just the top of the seat above water. For some hours we waited for the storm to abate, but the longer we waited the higher the water rose.

—Percy Megargel, on an attempt to ford a Wyoming stream

In 1905 two Oldsmobiles became the first automobiles to race across the country. A U.S. Department of Agriculture agent organized the contest, making it among the federal government's most visible early efforts to promote good roads for autos.

On May 8, 1905, two 1904 Olds runabouts started from New York City, in hopes of breaking the 33-day transcontinental record that a 4-cylinder, air-cooled Franklin auto had set in 1904. At the very least, the Oldsmobile crews wanted to reach Portland, Oregon, within 44 days for the June 21 start of the annual convention of the National Good Roads Association, held in conjunction with that city's Lewis and Clark Centennial Exposition. Olds Motor Works would pay all expenses and offered $1,000 to the winner; the loser would keep his car.

The company selected Dwight B. Huss and Olds factory mechanic Milford Wigle to drive the car dubbed "Old Scout," and Percy F. Megargel and Olds mechanic Barton "Bart" Stanchfield to drive "Old Steady." (See Figs. 1.1 and 1.2.) Newspapers often had trouble spelling the unfamiliar names. One *Cheyenne Daily Leader* reporter, possibly assisted by a puzzled type-setter, managed to misspell all four names in one artfully crafted sentence referring to Columbus, Nebraska, "since leaving which place Messrs Huff [sic] and Wengle [sic] have not seen Percy Mergarel [sic] and Bart Stanfield [sic], who are the crew of Old Steady."[1]

Minus their fenders but otherwise in stock condition,[2] "the two little 7-horsepower runabouts [were] fresh from the stock room and as alike as two peas," *Automobile* declared.[3] As 1904 curved-dash Oldsmobiles, the autos had tillers—steering wheels replaced tillers on the 1905 models—and a top speed of nearly 30 mph. One-cylinder engines were situated under the body of the Olds cars, machines as simple and reliable as this troubleshooting advice from an early Olds owner's manual: "Always consider that it has run and will run again if conditions are [the] same."[4] Fitted with 28-inch wheels, the 800-pound autos were only slightly improved from the Olds runabout in which Lester L. Whitman and Eugene I. Hammond in 1903 had made the third-ever transcontinental auto crossing and the first in a car so light.[5]

Fig. 1.1 Dwight B. Huss, left, and Milford Wigle in Old Scout. (AAMA)

Fig. 1.2 Percy F. Megargel, left, and Barton Stanchfield in Old Steady. (AAMA)

The Olds earned an international reputation for staunchness in late 1904: Maurice Fournier of Paris and his chauffeur drove an Olds runabout on a one-month, 3,000-mile tour of Europe, winning a $20,000 bet "to prove that the Oldsmobile runabout was something more than a car for city streets and well kept parks."Olds Motor Works advertised its reasons for selecting its 7-horsepower runabout for the 1905 coast-to-coast trip: "Best hill climber ever built. Has fuel capacity for over 100 miles of travel. Control is by one lever and two brakes. Nothing to confuse. It is immensely strong. Breakages are rare and there is nothing to watch but the road."[6]

Man Against Nature

The 1905 U.S. crossing—variously estimated at 3,500 to 4,000 miles, for the Oldsmobiles had no odometers—traversed New York, northwestern Pennsylvania, Ohio, Indiana, Illinois, Iowa, Nebraska, a few miles through northeastern Colorado, Wyoming, Idaho and Oregon. Heavy rains and mud cemented a cooperative bond between the competing crews during the first half of the trip. A true race developed only west of Omaha, Nebraska, but the entire trip was as much a battle of man against nature as a race between cars.

The topless, wood-bodied Olds runabouts wallowed through hundreds of miles of deep mud and sand and forded scores of streams and rivers. It took a day and a half to repair Old Steady after a Nebraska bridge collapsed. Old Scout ran away on a steep Idaho mountain road, crashed and needed a similar overhauling; it also nearly disappeared in a Wyoming sinkhole. The drivers, who were often lost on New York state's unmarked roads, would waste a half day at a time following dead-end trails in "the vast and unfrequented stretches of the far West— the *terra incognita* of the American automobilist," as *Automobile* termed it.[7] All four men got sick but nonetheless met the challenge of deep sand, unbridged rivers, kicking horses, hordes of true road hogs in Iowa, and weather that ranged from rain and snow to desert heat.

A Crying Need for Good Roads

The race idea came primarily from James W. Abbott, special agent for the Rocky Mountains and Pacific Coast in the U.S. Department of Agriculture's Office of Public Road Inquiry. He took the post "a few months" before the November 1900 organization of the National Good Roads Association, Abbott told the *Portland Morning Oregonian*.[8] Abbott had been involved with an earlier transcontinental crossing. On July 11, 1903, in Price, Utah, Abbott met with the crew of the second automobile to cross the country, "his interest in the expedition being due to the bearing it might have on the good roads movement in the West."[9]

Other sources cite different origins for the race, however. According to the *Chicago Daily News*, the National Association of Automobile Manufacturers instigated the contest "with a view to testing the feasibility of a national highway." The race grew from a conference between Abbott and Roy D. Chapin, Oldsmobile sales manager, in his capacity as chairman of the Good Roads Committee of the National Association of Automobile Manufacturers, "and is undertaken to show the crying need for good roads," according to the *New York Tribune*.[10] Chapin's involvement explains Olds Motor Works' participation in the coast-to-coast race. As the *Chicago Inter Ocean* describes the conversation between Abbott and Chapin:

> They were talking about the coming national good roads convention to be held at Portland from June 21 to 24, and thought that it would stimulate an interest in the question if an automobile trip should be made. As a result the Oldsmobile people offered a prize to the two couples making the trip in their cars.[11]

Though Oldsmobiles became the first cars to take part in a transcontinental auto race, the idea itself was hardly new. Newspaper publisher William Randolph Hearst had a similar idea nearly two years earlier. In autumn 1903, Hearst offered a $1,000 cup to the winner and "a valuable medal" to all participants in a transcontinental race tentatively scheduled to start in the spring of 1904. "From fifteen to fifty competitors are expected to take part, some assurances having already been gained," *Automobile* reported. "Some of the more prominent European road drivers will be induced to enter, making the affair of international importance." Cars could start from any point on either coast and would be placed in classes according to weight. For whatever reason, the Hearst transcontinental race—another of his attention-grabbing newspaper stunts—never came to pass.[12]

Popularizing Long-Distance Trips

Abbott "believes that this transcontinental tour will do a great deal toward securing better roads for many of the Western States," *the New York Times* said regarding the 1905 race. "In California there are many active automobile centres, but in the far Northwest the number of machines is yet very few," Abbott told the newspaper. "The good roads movement is absolutely necessary to make the automobile more popular, and if the two cars now starting succeed in making the trip on time, which will be within forty-four days, I think it will pave the way for future trips across the continent by many private motor car owners who will undertake the journey, perhaps, more leisurely, for pleasure."[13]

Abbott preceded the cars by train to plot the specific route based on whatever scraps of information and local gossip he could gather. He was evidently on hand at every overnight stop made by the leader in the race. In 1905, the best roads were railroads, so Abbott had already decided that the cars would follow the general route of the Union Pacific Railroad from Omaha to Granger, Wyoming, and then parallel the Oregon Short Line to near the Idaho-Oregon border, where they would strike out on their own.

Helpful directions from railroad officers who as yet perceived no competitive threat from the automobile "were important if not absolutely controlling factors in the success of the recent contest between Old Scout and Old Steady," Abbott declared afterward. He started the cars as late as possible in the spring of 1905 to avoid the lingering snows in Oregon's Cascade Range, and assumed a pace of 80 miles daily "so as to safely cover all the contingencies of a trip about which there were so many uncertain factors."[14]

Getting Started

The automobile journals announced the cross-country race in late April. The Olds factory chose drivers Huss and Megargel from "several hundred"[15] applicants only three days before the race began in New York City on May 8. Consequently, the 9:30 a.m. start from Harrolds Motor Car Company, an Oldsmobile and Pierce Arrow agency at 59th and Broadway, was without fanfare. (See Fig. 1.3.) Photos reveal that a few dozen spectators witnessed the start, a far cry from the thousands who packed Times Square for the start of the New York-Paris race three years later. The intersection was and still is known as Columbus Circle for a statue of Christopher Columbus, who—like the 1905 automobile explorers—faced unknown dangers in his westward trek.

"A small group of tradesmen, newspaper men and photographers watched the start of the two cars from Columbus Circle, and a number of persons, in several other cars, accompanied the transcontinental tourists as far as Central Bridge, over the Harlem river," *Automobile* wrote.

Fig. 1.3 The two curved-dash Oldsmobiles at the start of the race. The column, background right, is a memorial to Christopher Columbus. (OHC)

Fred Wagner of the New York Athletic Club "gave the official word to go at 9:30 o'clock, and accompanied by a dozen automobiles, they were soon lost to sight up Eighth Avenue," *a New York Times* reporter observed.[16]

When the cars reached the Harlem River bridge at 9:55 a.m., accompanying newsmen seized the opportunity to stage a photo, according to *Motor World*. "There the newspaper men and well wishers who accompanied the tourists assumed attitudes about the two cars, held their hats aloft and kept their mouths open long enough for the photographers to snap the scene of the 'cheering multitude.'"[17] (See Fig. 1.4.)

At the start, both cars carried signs on each side in script lettering: "Oldsmobile/Enroute/New York to Portland Oregon/$1000 Race." On the top of the curved dashes, nameplates in a similar script hand identified the autos as "Old Scout" and "Old Steady." But photographs show that between Omaha and Laramie, Wyoming, Old Scout received new signs and a new, larger nameplate, both lettered in a bold, sans serif hand. Old Steady got similar new signs—

Fig. 1.4 A staged farewell celebration at New York's Harlem River bridge. (AAMA)

presumably at the same point on the trail, though neither Huss nor Megargel mentions the transformation or the reason for it. Olds Motor Works perhaps wanted more visible signs on the cars. The new signs correctly spelled "En Route" as two words but they incorrectly spelled the cars' destination as "Oregeon."

"The travelers started lightly equipped, carrying 200 pounds of luggage, tools and parts in a steamer trunk strapped at the back of the seat on each car," *Automobile* said. "One spare tire is carried on each machine, and each is fitted with a swivel searchlight bracketed on the left side … These lights are so hung that they can be reversed and directed onto the engine in the rear of the body." The searchlights were mounted on the left side of Old Steady's dashboard and the left side of Old Scout's seat. The autoists took a cue from the previous Oldsmobile transcontinentalists, Lester L. Whitman and Eugene I. Hammond, who wore their clothes until they became hopelessly soiled, and then bought new ones. "The drivers carry only the clothes they wear and their leather jackets," as the *New York Tribune* described it. "Their wardrobe will be replenished whenever necessary, the old garments being thrown away."[18]

Three Mistakes

Afterward, organizer Abbott acknowledged three mistakes, or "incidents of hasty preparation and naturally limited storage capacity." One was placing an acetylene tank for the headlamp on the left side of each car, "where it always tended to weigh that side down excessively.… Another mistake was in not carrying at all times sufficient suitable clothing for protection against severe climate, which may be encountered at any season of the year in a transcontinental trip."

The worst mistake was packing a box—perhaps meaning the steamer trunk on each auto—with loose tools and parts, Abbott declared. "It was a cause of breakage and delay from beginning to end of the race. Concentrating material of small bulk but heavy weight in an enclosure of rigid shape inevitably makes a battering ram which is a constant menace to springs, frame, and all connected mechanism." It is better to wrap small bundles of tools and parts in sacks and store them throughout the car to distribute their weight, he said.[19]

Published reports noted just one difference between the two runabouts: Old Scout used Fisk while Old Steady used Diamond tires.[20] On the third day of the race, "the engine of Old Scout seemed to run a little better than mine, and Huss steadily drew away,"[21] said Megargel, who perhaps drew the wrong conclusion. As he later revealed, Old Steady was the heavier car because Megargel and Stanchfield outweighed Huss and Wigle by 40 pounds.

"Don't drive your 'Oldsmobile' 100 miles the first day," the Oldsmobile owner's manual warned. "You wouldn't drive a green horse 10 miles till you were acquainted with him. Do you know more about a gasoline motor than you do about a horse?"[22] With two mechanics on board, the motorists could honestly answer "yes" to that question; they traveled the nearly 200 miles to Albany the first day.

Both mechanics worked for Oldsmobile: Old Scout's Milford Wigle for the Detroit factory and Old Steady's Barton Stanchfield for the Lansing, Michigan, factory, "between which there is said to be the keenest rivalry, so that each party can be relied upon to do everything that is fair to finish the tour first,"[23] according to *Automobile*. Though Huss and Megargel are remembered as the "drivers" on the first transcontinental race, the mechanics also drove, Megargel said.

The Magnificent Men in their Motor Machines

Wigle had plenty of experience behind the tiller. He drove one of four Oldsmobiles entered in a 29-car, 100-mile Chicago endurance contest on August 2, 1902. Wigle finished second among the Olds contingent, 22 minutes behind Roy D. Chapin, future president of Hudson Motor Car Company, who was then primarily an Olds Motor Works test driver. Driving Oldsmobiles in various races between 1902 and 1904, Wigle won three gold medals and 12 silver cups for the automaker, according to George S. May in *R.E. Olds: Auto Industry Pioneer*.[24]

Megargel was an avid bicyclist who later pursued an interest in autos. Once a central New York representative to the National Cycling Association, Megargel edited and published a bicycling magazine called *Side Paths* at Rochester, New York, from 1897 to 1902, according to a press account.[25] Sewall K. Crocker, who with Dr. H. Nelson Jackson made a 1903 coast-to-coast trip in a Winton auto, and Eugene I. Hammond, who accompanied Lester L. Whitman on Whitman's 1903 and 1910 crossings, had also been professional bicycle riders. Thus former bicycle racers took part in at least three of the first five transcontinental automobile crossings.

Megargel in 1904 drove a 1-cylinder, 1,200-pound Elmore tonneau pathfinding car in blazing the trail for the American Automobile Association's 1,200-mile endurance run from New York City to the World's Fair at St. Louis.[26] He returned by auto to New York City in time to drive the same two-cycle Elmore to St. Louis with other autoists on the tour. During the contest, a high spot in the road near Rochester, New York, tore away the Elmore's brake rods; shortly afterward, Megargel smacked into the back of a Pierce Great Arrow, slightly crushing the Elmore's radiator.[27] Megargel, "a graduate of the military school at Aurora, N.Y.," who "has long been identified with newspaper work in Rochester," was advertising manager for the E.R. Thomas Motor Company, makers of the Thomas car in Buffalo, New York, during the six months preceding the 1905 race to Portland.[28]

A Detroiter who grew up in Clyde, Ohio, Huss "started out as a bicycle racer, like Barney Oldfield," he explained in a 1954 *Life* magazine article. "First race I was ever in was a bike race from Clyde to Green Springs, Ohio, in 1890. The prize was a grand piano. I was ahead when a chicken flew into my drive chain and got wrapped around the sprocket. Lost the race. I quit [bicycle] racing a long time ago, myself." Later, Huss "won a gold medal in the English Reliability Trials in 1903, driving a standard Olds runabout, and during the last two seasons drove the Oldsmobile racers with success on the track," *Automobile* said in a 1905 article.[29]

"In 1903 as representative of the Olds Motor works, Mr. Huss took the first American cars to Europe and within a year had established agencies in every major country of the continent," a retrospective article contends. During the period, he "entered and won numerous European auto races," as well.[30] Huss drove an Oldsmobile tonneau in the same New York–St. Louis run in which Megargel participated. One of his adventures occurred on muddy roads west of Utica, New York:

> Mr. Huss, Oldsmobile, was among those who experienced bad skids during the run to Syracuse. His driving chain had broken while coming down a steep hill, and he had applied the hub brakes so quickly that the rear wheels locked and the car slid down into a gulley [sic]. Luckily it was not injured, and after the chain was fixed he was able to proceed. Mr. Huss was also troubled by his low-speed gear getting out of adjustment so that it would not engage properly, and he was consequently obliged to rush all the hills encountered and make them on the high gear.[31]

Trails and Trials of New York

Rules for the 1905 transcontinental race required the crews to obey local speed laws and remain with their cars. "One man will not go on ahead by rail, picking up the car at some future point," the *New York Times* said. "If the men wish to sleep in a more comfortable bed than can be afforded by the light runabouts, they must do so at their own loss of time.[32] The $1,000 prize—which in 1905 was twice the average annual wage[33]—"is sufficiently large to induce the men to put forth every energy to bring out every ounce of endurance and effort of which the machines are capable," *Automobile* noted.[34]

Olds Motor Works also offered amateur and professional photographers a total of $150 in prizes for the best photos of the cars and drivers en route. Arrangements were made for Standard Oil Company to ship gasoline and oil to Western points on the route. Gas cost as little as 15 cents a gallon in Ohio. Upon paying 60 cents a gallon at Cokeville, Wyoming, Megargel believed himself "fortunate that the little runabout is economical of fuel."[35]

Rain, mud, skittish horses and unmarked, unfamiliar roads were among the biggest trials of the trip from New York City to Omaha. The larger cities on the first leg of the trip included Albany, Utica, Syracuse, Rochester and Buffalo in New York; Erie, Pennsylvania; Cleveland and Toledo in Ohio; Chicago; and Davenport and Des Moines in Iowa.

"Exceptionally fine weather and good roads"[36] allowed the racers to reach Albany their first day. But an overnight rain left the roads "full of mud and water," Megargel said. "Owing to the mud guards [fenders] having been stripped from both Old Steady and Old Scout, we were soon covered from head to foot with mire, while our baggage and supplies, strapped on the rear deck, were almost unrecognizable when Syracuse was finally reached" at the end of Day 2. Despite their mud covering, "All four overlanders are so sunburned already that they can scarcely wear collars or gloves,"[37] Megargel wrote that night to *Automobile*, which ran his accounts as a regular feature titled "Diary of the Transcontinental Race." In addition, both Huss and Megargel wrote of their experiences in *From Hell Gate to Portland*, a 46-page Olds post-race booklet.

Even in New York the racers often lost their way. Between Syracuse and Rochester on Day 3, the cars got lost on different routes through marshes near Montezuma. They arrived in Rochester simultaneously, however, and the crews ate lunch together, Megargel writes. Starting from Buffalo the next morning,

> As usual, both crews lost themselves and traveled many more miles than were necessary. In fact, after making numerous turns, Old Steady struck a section of the country where automobiles appeared to be unknown. A small boy on his way to school, whom we stopped to ask about the road, was frightened and commenced to cry violently. Soon after, a girl, after taking one look at the begoggled, mud covered tourists, gave a yell and started down the road at top speed. When she saw we were overtaking her she made for the side fields, falling several times before she finally convinced herself we were not really after her. The next youth we met dived bodily through a barbed wire fence and streaked it across the fields at top speed.[38]

"A Human and Mechanical Mud Heap"

The drive from Buffalo to Cleveland was through deep mud. In particular, rain left Ohio roads between Conneaut and Cleveland in "terrible shape—low speed a large part of the way and numerous mud puddles of uncertain depth to ford," Megargel said. "At one point we ran for a quarter of a mile without seeing the road at all, the water in some places coming above our axles."[39]

The cars left Buffalo at about 7 a.m. Thursday, May 11, and at 1 p.m. stopped in Erie, Pennsylvania, "just long enough to get dinner and to fill their gasoline tanks," reported the *Erie Dispatch*. "Murphy Bros., who have charge of the local Oldsmobile garage, sent another car to meet the tourists at North East and pilot them into the city. Huss and Megargel report exceedingly bad roads owing to the wet weather, and they say they have great difficulty in keeping from slipping off the road when going at a high rate of speed." The Oldsmobiles were holding up well, however, and "although completely covered with mud, everything was in first class condition and the drivers have had no trouble except with the tires."[40]

Old Steady arrived in Erie with a punctured tire, Megargel explained. Old Scout stopped Thursday night at Painesville, Ohio, short of its Cleveland destination, because "the occupants of the car were completely worn out," as the *Cleveland Plain Dealer* put it. "The trip from Buffalo to Painesville was a heartbreaking one for the two men and was made under the worst possible conditions. Shortly after the start from Buffalo ... the Old Scout was caught in the big rainstorm."[41]

The next day, Friday, May 12, after a drive of 1 hour, 50 minutes, Old Scout reached a Cleveland garage at 7:30 a.m., "a human and mechanical mud heap, but animated to a degree surprising, considering the fact that they had come through the worst roads imaginable," the *Cleveland Leader* observed. Ralph R. Owen—brother of Ray M. Owen, the Olds distributor for New Jersey, New York and Ohio—"and a party of automobile enthusiasts and newspaper men" met Huss and Wigle on Euclid Avenue and piloted them to the garage.

> The car showed the effects of the struggle through the mud and water, too, being covered to a depth of several inches with dried clay. The men were covered as deeply, but happiness was their share, for they were far in the lead.…

> The car was cleaned at the garage and preparations made for the start to Toledo, which city will be reached Friday evening if it is possible to accomplish the feat, and Huss, the famous Olds driver, believes it can be done.[42]

Old Scout's modest lead soon evaporated. Megargel and Stanchfield in Old Steady arrived in Cleveland from Conneaut—where they had stopped the night before "to renew a short-circuited battery"[43]—at 10:10 a.m. Friday. Both cars started for Toledo at 12:35 p.m., the *Leader* said. "But the roads were so muddy that time was lost and the best the tourists could do was to reach Clyde, where they remained over night," according to the *Toledo Blade*.[44] It reported that Old Steady had a 30-minute lead heading into Clyde. The two crews had decided to remain in Clyde, "Huss desiring to visit his family and Megargel wishing to stop at the Elmore factory," recounted *Cycle and Automobile Trade Journal*.[45] It was an Elmore car that Megargel drove in the 1904 New York–St. Louis endurance run.

Rather Slow Time

The cars reached Toledo on Saturday, May 13, reported Saturday's *Toledo Blade*:

> Covered with dirt, but happy and feeling good, four transcontinental tourists arrived in Toledo in their autos this morning, remained a few minutes to oil their machines, and hurried on their trip to the Pacific coast.…

> Both cars were sent out of Clyde this morning at 6 o'clock and the "Old Steady" arrived in Toledo at 8:45. Its running mate was delayed halfway between Clyde and Toledo by a puncture and did not arrive until 9:55, giving its opponent an hour and ten minutes start out of Toledo. The machines will be driven at record speed today, as the autoists expect to reach Chicago by evening.[46]

The Toledo newspaper perhaps confused the two cars. According to Megargel, Old Scout reached Toledo first. After repairing a puncture between Clyde and Toledo, Megargel and Stanchfield in Old Steady sped after the leaders. "Just before entering Butler, Indiana, we rounded a corner and came upon Huss and Wigle," Megargel recalled. The roads had worsened between Toledo and Waterloo, Indiana, however, Megargel said, and rain also fell in northeastern Indiana, where the two crews ate their evening meal together at Kendallville. "It began to rain hard again before supper was finished, and as it was nearly 10 o'clock we decided to remain all night."[47]

Megargel's daily reports for *Automobile* magazine contain no entry for Sunday, May 14; the *Chicago Daily News* said the cars stopped Sunday night in different Indiana towns—Old Scout in Hammond and Old Steady in La Porte. They reached Chicago, 1,150 miles from New York City by their route, on Monday, May 15, a week from the start in New York City. "Generally speaking," *Motor World* said in summarizing the trip to Chicago, "where they have not wallowed in mud, they have either waded through water or been drenched by it."[48]

"We have made the run from New York to Chicago in rather slow time on account of the rains and storms farther east, but we can expect to have some exciting adventures when we get out in the mountains," Huss told the *Daily News* upon arriving first in Chicago at between 6:30 and 7 a.m., according to conflicting accounts. The men drove to the Oldsmobile garage to have Old Scout cleaned and adjusted.

> At the resting place in Michigan avenue Huss and Wigle had much to say of their trip across New York, Pennsylvania, Ohio and Indiana. The cars were run at about eighteen or twenty miles an hour and the best part of the trip was made in the last three days. In New York from Albany, the first night's stopping place, to Syracuse continuous rain and heavy winds kept down the speed and made riding extremely difficult. Several times the cars were stopped by mud and the tourists were dragged to dry ground by lenient farmers, who showed their interest in the trip by [giving] frequent breakfasts and dinners to the tired travelers.[49]

Megargel and Stanchfield were late into Chicago, arriving at noon Monday, some five hours behind the leaders, according to the *Daily News*, *Inter Ocean* and the telegram Abbott sent to the Olds factory in Lansing, Michigan. In his *Automobile* diary, however, Megargel contends he reached Chicago just two hours behind Huss. Regardless of its length, the delay was due to Old Steady's engine, which had been misfiring since Saturday afternoon. "We overhauled one thing after another in an effort to locate the trouble," Megargel said. "First it was the carbureter, next the oil, then the batteries, spark plugs, valves and the hundred and one things that might cause a missing of any gasoline car."[50] Near Chicago they found the cause—a poorly connected ground wire.

Sticky, Black Illinois Mud

Despite receiving frequent drenchings and averaging just four hours' sleep nightly during the previous week, the crews were in excellent spirits upon reaching Chicago, Megargel said. "I think I have gained about three pounds on the trip thus far," he wrote several days later from eastern Iowa. "If this increase keeps on we will be sadly handicapped by the time Portland is reached."[51] As far as Chicago, the route was similar to the one Huss and Megargel followed during the AAA's 1904 New York–St. Louis tour. With an "imperfect knowledge" of the roads ahead, the cars left Chicago 30 minutes apart at mid-afternoon Monday, May 15, hoping to reach Davenport, Iowa, in two days.[52]

Then more rain hit. (See Fig. 1.5.) It started late Monday afternoon and continued all day Tuesday, which made Wednesday, May 17, "the shortest day's mileage since leaving New York and the hardest work," Megargel wrote from the overnight stop in Geneseo, Illinois.

> Illinois mud, even when taken at a racing gait, brought our mileage down to 64 miles for the day, and we worked tooth and nail ever since 5 a.m. It is now 10:30 p.m., and we have just arrived in town.…
>
> The mud all day has been that sticky, thick, black mud that only Illinois and Missouri can boast of. To get through it we were obliged[,] on several occasions, to use blocks and tackle. The water in our tank boiled, but the engine never faltered, although every one of the four wheels was so stuck up with mud as to render the spokes invisible.[53]

Huss and Wigle were hub-deep in mud for five miles between Chicago and Aurora. At 7:30 p.m. Tuesday, May 16, Old Scout reached Mendota, Illinois, where it spent the night. "Roads

Fig. 1.5 From Indiana westward, the Oldsmobiles churned through miles of sticky Mid-western mud. A crewman snapped this photo at an unrecorded location. (OHC)

almost impassable last thirty five miles," Huss wired the factory from Mendota. "Record run dubious," Megargel observed in his nightly telegram from the same town. According to *Cycle and Automobile Trade Journal*, Old Scout kept its lead to beat Old Steady into Geneseo on Wednesday. But,

> Megargel overtook Huss on the way to Davenport, which was reached May 18 at about 2:30 p.m. All four men were covered with mud from head to foot. They had arrived at Moline about 2 o'clock, and were met there by a large delegation of motorists. They were escorted across the Mississippi with many cameras pointed at them as they proceeded. Two hours were spent in cleaning the cars at Davenport, and when they left that city they had an escort of about a dozen cars.[54]

Roads Like Russia's

To reach Davenport on Thursday, May 18, the two runabouts splashed through Rock River floodwater near Moline, Illinois, so deep that it flowed over Old Steady's floorboards. (See Fig. 1.6.) Only an under-car apron kept the carburetor dry and the engine from stalling. The mud in Illinois, Iowa and Nebraska, was "soft and sticky beyond description," Huss wrote. "We rode in mud up to the hubs at times, and I have seen the wheels many a time with so much

Fig. 1.6 Huss pilots Old Scout through floodwaters as deep as Megargel described. This unidentified scene most likely shows the Rock River but may depict Iowa flooding. (OHC)

'Gumbo' sticking to them that they would weigh 250 to 300 pounds each."[55] Both cars spent Thursday night some 30 miles west of Davenport in Wilton, Iowa. They again traveled muddy roads during the next day's drive from Wilton to Brooklyn, Iowa. "One motherly old woman who passed us as we were working on our car by the roadside exclaimed: 'You poor fellows, I'm awfully sorry for you,'"[56] Megargel recalled.

Route-planner James W. Abbott had made "diligent inquiries" in Chicago about the best route to Davenport. "We concluded after the boys had met that horrible experience in the mud east of Moline that we had not been sufficiently informed," he said later. To avoid the same mistake, "at Davenport and Des Moines we made a most drastic search" and found a route that, though indirect, would take the cars safely across the rampaging South Skunk River at Reasnor, Iowa, near Des Moines. Megargel and Huss, however, ignored Abbott's advice by following the more direct route sketched by a liveryman in Grinnell, setting up on Saturday, May 20, "that harrowing scene which transpired on the Skunk River bottom.... Near Newton [they] plunged without warning into a watery abyss from which their escape was really miraculous," as Abbott described it.[57]

"Huss and I agreed to help each other across, or neither of us would ever have reached the other side," Megargel said (see Figs. 1.7, 1.8 and 1.9):

Fig. 1.7 A surprisingly cheerful Huss, left, and Wigle attempt to extricate Old Scout from Skunk River floodwaters before Old Steady (behind them) ventures ahead. (OHC)

First, one car would plunge into the water and mud until stuck, then we would rig our blocks and tackle, attach it to the car axle and bring the other end back to the car still on dry land. By all of us pushing, and the free car towing, we managed to work our way across a swamp that every one told us was impassable.

All of us worked in water and mud nearly waist deep.... The water was so deep that at one time all of the four wheels were out of sight, and a stream flowed freely through the flooring of the wagon, soaking everything, filling the carbureter with water and short-circuiting the battery.[58]

Fig. 1.8 Wigle, left, and Huss in another view of the stranded runabout. (OHC)

Still, the cars covered 72 miles that day, but arrived in Des Moines early in the evening instead of early in the afternoon, as planned. "Huss, who has toured Europe, says the roads of Russia are the only ones that compare in vileness with those of Iowa," *Motor World* reported.[59]

Megargel and Stanchfield spent most of Sunday, May 21, "overhauling" their car in Des Moines. They left about 4 p.m. and, along with Huss and Wigle, spent the night at Dexter, some 54 miles west of Des Moines, according to a Megargel telegram.

Genuine Road Hogs

Though generally tamer in the East, horses were a constant menace to the small cars, Megargel said. Scared by the "two sputtering cars," a horse bolted while pulling three women in a carriage near Earlville, Illinois. "Milford Wigle, Huss' mechanician, jumped out of the car

Fig. 1.9 Stuck again, a less-cheerful Huss ponders his predicament. (OHC)

and by a desperate sprint … managed to grasp the runaway's bridle. He was dragged some distance, but retained his hold and brought the frightened animal to a stand," Megargel recounted. "In passing horses on the road we have to be extremely careful, as several have kicked most viciously at our car, and its occupants," he wrote from Iowa.

Free-roaming pigs were an Iowa hazard. "There are hundreds—yes, thousands—of pigs on every farm, I believe. These droves of ham and pork chops occasionally block the road so that we are obliged to slacken speed and blow the horn for a long distance before we can get a clear road." One consolation was "that ill feeling against automobilists that one encounters so frequently among the farmers in the East is lacking out here, and every farmer met with is your friend, no matter how badly you scare his horses or how much damage his nag does before you finally get past him on the road."[60]

According to Huss, Iowa fielded larger crowds of spectators than any other state on the race route. In the Hawkeye State,

> the farmers drove in for miles to be present as we went through some village.… In the great open spaces, the folks looked on the automobile differently than did those in the city. In the East the folks regarded it largely from a pleasure standpoint; in the West, from the utilitarian. If it was to be worthwhile for them it had to be improved transportation, not a playtoy. So their interest in the race was not sporting, but downright practical.[61]

Midway in Omaha

Fifteen days of tortuous travel left the cars and their pilots "showing the effects of a hard struggle with mud, water and rough roads" upon their Monday, May 22, arrival in Omaha, nearly the halfway point. "It is doubtful if any transcontinental wagon outfit since the earliest forties [1840s], ever attracted as much attention on the streets of this city," Megargel wrote.[62]

Olds Motor Works and its dealers made the most of the free publicity attending the race. While the cars were in Omaha, the local Powell Automobile Company asked in an ad:

> Why is the car [Oldsmobile runabout] able to stand up so well under the fearful rack and strain put upon it? Because every pound in its make-up is a pound of endurance. Because it is simple, durable and easily handled. Because it is not an experiment, but the finished result of years of experience.... Because—briefly—the car is built to run and does it, whether on the city boulevards or over the roughest mountain passes.[63]

Route maps "have been placed on exhibition at all of the Oldsmobile agencies throughout the country, and the tourists are to call at the agency in each of the cities through which they pass," *Automobile* reported.[64] Old Scout and Old Steady differ "in no particular from the thousands of these cars now in daily use," Olds Motor Works said after the contest. "When you invest $650 in a car which can successfully stand the strain of a 3,500-mile race across the American continent, under all kinds of road and weather conditions, you will get pretty satisfactory returns for daily use."[65]

Wilder Country Ahead

During 2½ days spent in Omaha, the crews heard that "from Boise City [Idaho] to Salem, Oregon, civilization drops from sight and with the exception of an occasional ranch, we shall meet neither people nor living objects other than the wild animals that are supposed to roam at large over this territory," Megargel said.[66] Both crews bought guns, though Megargel proved a better driver than a marksman. He spent 30 minutes with his .38-caliber revolver and wasted a box of cartridges in a vain attempt to shoot a telegraph pole from across the road. Huss had a better aim. (See Fig. 1.10.) For several days previously, Megargel would write from Rawlins, Wyoming, "we have been wearing our revolvers in plain sight, with our belts bristling with cartridges, as do many riders we meet along the road."[67]

The cars left New York City with about 200 pounds of baggage apiece, Huss noted. "At Omaha we equipped for the wilder country ahead, taking on extra tanks for desert driving, sand tires, fire arms and cooking utensils, all of which increased the weight materially, so that when the machines left Omaha each had a load of practically four full grown persons."[68] The extra gasoline tanks would allow "for runs of 200 and 300 miles between sources of supply," *Motor World* said.[69] The men replaced the steamer trunks with taller wooden boxes that fit neatly between the rear luggage rails and the new behind-the-seat gas tanks. Photos show that Old Scout—and presumably Old Steady, too—carried a shovel in the West.

Fig. 1.10 Huss and his quarry in a photo that Wigle snapped somewhere between western Wyoming and Portland. (OHC)

Both cars were supposed to start westward from Omaha late in the afternoon of Wednesday, May 24, Day 17, but because of a hard rain the travelers postponed the start to the following morning, according to Megargel. During the last full day that they would travel together (See Fig. 1.11.), Old Scout and Old Steady left Omaha for Fremont on Thursday, May 25. Their route beyond Fremont would take them through Columbus, Grand Island, Kearney, Cozad, North Platte and Sidney in Nebraska; and Cheyenne, Laramie, Rawlins, Rock Springs, Granger, Kemmerer and Cokeville in Wyoming. From the Nebraska panhandle they also dipped a few miles south across the border to Julesburg, Colorado, following the established trail alongside the Union Pacific Railroad. (See Fig. 1.12.)

Both cars reached Fremont about 10 p.m. Thursday. "High water and bridges out," Megargel wired the factory. "They are now two days behind time, having made slow progress the last few days on account of bad roads," Friday's *Fremont Evening Tribune* reported. "Early this morning they started on west. It was the intention to reach North Platte this evening."[70] Muddy roads thwarted that intention, however.

Fig. 1.11 Old Steady makes a repair stop on a drying road in eastern Nebraska, where the racers traveled together for the last time. Sand tires strapped to the front of the cars fore-tell the rigors ahead. (OHC)

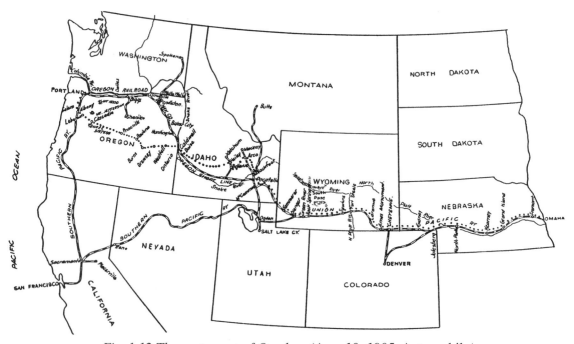

Fig. 1.12 The route west of Omaha. (Aug. 10, 1905, Automobile*)*

Bridge Breaks; So Does Old Steady

Though driving separately, both cars had made the same westward progress by Friday evening, May 26. Old Scout stopped for the night in Grand Island[71] and Old Steady spent the night due north of Grand Island in Palmer.

But next morning en route to Grand Island, "we were unfortunate enough to crash through a bridge, part of which had been washed away," Megargel wrote. Pitched into the water, the men "had a narrow escape from drowning," *Motor World* asserted.[72] Megargel, however, mentioned only the damage to Old Steady: "The fall broke both side springs of our car, the strut rod on the front axle, and bent both the front and rear axles. Stanchfield succeeded in making temporary repairs which enabled us to travel some fifteen miles into town, where much valuable time was lost getting Old Steady in shape again."[73]

The accident cost them a day—they were able to leave Grand Island at daybreak on Sunday, May 28—but Megargel later contended that the Nebraska mishap actually cost them much more time. True, the damage "was mended, but would break again at the slightest provocation, and that meant the delay of half a day every time,"[74] he told the *Portland (Ore.) Evening Telegram*. The damage was not repaired permanently until reaching Boise, Idaho, according to Megargel.

At 5 a.m. Saturday, May 27, the day of Old Steady's bridge accident, Old Scout left Grand Island and pushed farther west, reaching Kearney at 3 p.m., the *Kearney Daily Hub* revealed. "A number of local automobilists met 'Old Scout' at Gibbon and accompanied it to this city.… The machine was thickly covered with mud when it reached Kearney and showed that it had not been passing through a drouth stricken country."[75] On Sunday, May 28, Huss and Wigle, on "roads [that] are in a terrible condition on account of repeated rains,"[76] spent all day traveling the 49 miles from Kearney to Cozad, passing through Overton and Lexington on the way.

Three miles west of Overton, "we got stuck in a bad mud hole," Huss wrote to *Automobile Topics*:

> And to make matters worse, it commenced to rain very hard. We walked two miles and got some shovels and boards, and succeeded in getting the machine out after three hours of hard work. We were literally soaked with mud, and both were badly chilled.
>
> When we reached Lexington we succeeded in purchasing some new clothes and hip boots, and started on. We had gone but three miles further when we slid into a ditch, and it took four men besides ourselves to get out.[77]

Old Steady, which left Grand Island early Sunday and traveled all day in rain, reached Kearney at about 6 p.m., according to the *Daily Hub*.

The two gentlemen in the car appeared to feel rather blue and did not take as kindly to the rain and mud as the two in "Old Scout." However, they pushed on at once, but were compelled to return to the city after going west for about six miles. The roads were too bad for them, so they remained in Kearney all night, starting out again early Monday morning.[78]

Pulling a Car or Pulling Our Leg?

Old Scout drove through Sidney in western Nebraska on Wednesday, May 31, about a day ahead of Old Steady, which arrived Thursday, the *Sidney Telegraph* reported.[79] Later in the day Wednesday, an "enthusiastic crowd of automobilists" met Huss and Wigle in Archer, Wyoming, to escort Old Scout the short distance to Cheyenne. "If you had seen that procession come into town you would have smiled. You would have seen Old Scout towing another car, besides carrying two passengers in each rig," Huss wrote. "We will not mention the name of the car, because it might cause some hard feelings."[80] But Huss did let slip that Old Scout later forded a river west of Laramie while its temporary traveling companion, a White steamer, got stuck.

It undoubtedly would have caused some hard feelings had Huss mentioned the make of the disabled car in Cheyenne. That's because, according to the *Cheyenne Daily Leader*, it was another Oldsmobile! The *Leader's* account and an independent story in the *Laramie Republican* also dispute Huss' contention that Old Scout was the tow vehicle. Further, both newspapers describe the "procession" as two autos, one of them in tow. Old Scout, which had left Julesburg, Colorado, 144 miles away, at 8:45 a.m., reached Sidney, Nebraska, at 12:35 p.m. and drove into Cheyenne at 10:30 p.m., "coughing like a cigarette fiend," said the next day's *Leader*:

> Accompanying "Old Scout" under its own power was one lone Oldsmobile driven by Thomas Myotte, the only one of the five machines which started for Archer yesterday afternoon to meet the racer to reach that destination. Trailing forlornly in the rear of Myotte's machine was a second Oldsmobile, attached to the leader by a cable of bailing wire. This machine was that of Dr. Crook, which became stalled while en route to Archer, and had to be hauled back to town. George Nagle's and Samuel Corson's Autocars and Dr. J.H. Conway's Oldsmobile, comprising the remainder of the reception parade, turned back before Archer was reached because of the poor roads....
>
> L.J. Ollier, traveling representative of the Olds Motor Works, met Old Scout at Archer last night at 9 o'clock, accompanying Thomas Myotte to that place, and escorted Messrs Huss and Wengel [sic] into the city.... Old Scout stopped in front of the Depot Cafe when it reached the city last night, and attracted much attention from passers-by to whom its record was known.[81]

James W. Abbott, who had taken a westbound train from Julesburg, was waiting in Cheyenne to greet Huss and Wigle, who "report the roads of Nebraska and Eastern Wyoming in miserable condition and say they experienced great difficulty in getting through," the *Daily Leader* said. "Since leaving Chicago they have had to contend with deep mud, and are almost two weeks behind their schedule." Another Cheyenne newspaper, the *Wyoming*

Semi-Weekly Tribune, reported, cryptically, "supplies have arrived here for the two motor cars and they will take on a stock at this point."[82] The *Tribune* did not elaborate.

Abbott was on solid ground, even if the two racers were not, in telling a reporter "that the present trip of the two Oldsmobiles will prove an incalculable value as it demonstrates so strikingly the need of good roads in the western country," as the *Wyoming Semi-Weekly Tribune* paraphrased it. "During the past two weeks the Oldsmobiles have lost almost a week on their schedule time as the result of rains, which placed the roads in deplorable condition and which shows that the ordinary rains, which may be expected each year, tend to render most western roads almost impassable."[83]

Old Steady "Almost a Wreck"

Old Scout left for Laramie at 3:30 p.m. Thursday, June 1, the day after arriving in Cheyenne. (See Fig. 1.13.) Huss and Wigle tried to leave early that morning, *Motor World* related. "'Old Scout' refused to budge when they turned the crank, and it was not until the coil was examined that they found the cause. The coil was found to be watersoaked and full of mud."[84]

Fig. 1.13 Huss poses by the Ames Monument, a landmark between Cheyenne and Laramie. The 60-foot-tall granite pyramid marks the highest point on the Union Pacific Railroad—8,247 feet. (OHC)

Old Steady, "sorely injured, but puffing gamely," reached Cheyenne at 9:15 p.m. Thursday, some six hours after Old Scout's departure and nearly a day behind the leader's arrival time. Old Steady was in no condition to go on immediately, however, according to Friday's *Cheyenne Daily Leader*. Megargel and Stanchfield had covered the more than 100 miles from Lodgepole, Nebraska, since 9 a.m. Thursday. According to the *Leader*, the men

> will lay over in Cheyenne today [Friday] in order that repairs may be made to "Old Steady." The machine crashed through a bridge near Columbus, Neb., and was almost demolished, but her plucky crew worked all night and managed to effect repairs which enabled them to reach Cheyenne under their own power. But "Old Steady" is almost a wreck, and must be extensively repaired. The accident at the bridge enabled "Old Scout" to attain its present long lead.[85]

Abbott, Ollier, and Charles E. Collins of the Diamond Rubber Company—the Akron, Ohio, makers of Old Steady's Diamond tires—met Old Steady in Cheyenne, some 2,300 miles west of New York City by the race route. Both cars would travel the 1,200 miles to Portland "in at most ten days, although the condition of the roads encountered will govern the time," the *Leader* erroneously predicted.[86]

Sometime after 8 p.m. Thursday, June 1, Huss and Wigle drove Old Scout into Laramie, where "the boys spent the night at the Johnson hotel,"[87] according to the *Laramie Republican*. (See Fig. 1.14.) By the following evening, Friday, June 2, "we got as far as Medicine

Fig. 1.14 A crowd that gathers as Old Scout leaves Laramie, Wyoming, includes James W. Abbott (facing left, partly obscuring the car's seat). Standing directly behind the car (head bowed, facing camera), Milford Wigle finishes packing. (OHC)

Bow at 7 o'clock, and just outside of the village we ran into the first sinkhole on the trip, and Old Scout dropped almost out of sight," Huss recalled. It took a block and tackle to extricate the car. On Saturday, June 3, Day 27, Old Scout's crewmen left Medicine Bow at 4:30 a.m., striking off from the railroad in hopes of making a long run. By 1:30 p.m., they had reached a small UP station just 35 miles away, Huss reported.

> I don't want to say much about the roads between these points, because it takes more than a pen to describe them. I don't know how Old Scout ever lived through it, but she did. We forded five big creeks [see Fig. 1.15] driving over rocks and stones and so much sage brush that Old Scout's radiator looked like a badly twisted piece of tin. At last we found the railroad, and we both swore to stick to the Union Pacific the rest of the trip. The everlasting sage brush was one of the worst nuisances we had to contend with.[88]

Old Scout "Forges" Ahead

By following the railroad, Old Scout reached Rawlins at 4 p.m. and kept going. According to *Motor World*, "After leaving Rawlins and while moving at a smart pace, in trying to dodge a big mud hole, 'Old Scout' ran into a bigger one; it pushed the [front] axle back eight inches, but did not break it, and Huss is taking chances until he can find a blacksmith shop."[89] He found one the next day, Sunday, June 4, upon reaching Green River, an over-night stop. Old Scout's crew discovered the blacksmith's shop at about midnight, Huss recalled. "Lots of persuasion and a $10 bill finally routed out the blacksmith[,] who lived above the shop.... The three of us took the axle off, started the forge going and got things fixed up in pretty good shape, reinforcing the axle with angle iron and clamps. Those homemade repairs, I might state, carried us through to Portland."[90] (See Fig. 1.16.)

Old Steady, which spent Friday undergoing repairs at Cheyenne's House Garage, left the city at 7 a.m. Saturday, June 3, for Laramie, some 50 miles away, according to Cheyenne newspapers. Driving northwest into the mountains, Old Steady reached Laramie at 2 or 2:30 p.m., according to conflicting newspaper reports, "travelling at a rapid speed and with its crew confident of overtaking 'Old Scout,' the leader in the race," the *Cheyenne Daily Leader* reported. Old Steady left Laramie at 3 p.m., Megargel wired to the factory, and spent the night at the Terry Fee Ranch. In a brief conversation with a *Leader* reporter, Megargel "stated that 'Old Steady' was in better condition than 'Old Scout,' and expressed confidence that his machine would overtake 'Old Scout'" before reaching Portland.[91]

He well may have, except more bad luck struck at Rawlins, an overnight stop: when Old Steady ran short of oil for the engine and transmission, the two men used "a small quantity of so-called gas-engine oil," bought at a store, Megargel said. He and Stanchfield left Rawlins early on Tuesday, June 6, Day 30, "at a good rate of speed, everything working finely, but when we had gone about five miles the car faltered and stopped.... When the two oils mixed they had formed a paint-like solution that the heat of the cylinder transformed into a hardened,

Fig. 1.15 Wigle dismounts when Old Scout stalls in an unidentified stream. (OHC)

Fig. 1.16 The men used clamps and metal strips, not angle iron, to repair the front axle. Here, Abbott, standing, meets with Huss, at the tiller, and Wigle late in the trip. A horseshoe adorns the car. (OHC)

sticky gum, and as the piston of an Oldsmobile fits the cylinder within 3-1000 of an inch, the coating of gummed oil soon stuck the piston."

Megargel walked back to Rawlins and hired a team of horses to retrieve the car. "We worked most of Tuesday night, and until 10 o'clock Wednesday morning cleaning out the engine. Gasoline, kerosene and lye had no effect, and it was not until we resorted to muriatic acid that the gum would dissolve. It was necessary to remove the body from the car and take the cylinder out."[92] (See Fig. 1.17.) Old Scout was $3^1/_2$ days ahead.

Lone John to the Rescue

Bad weather continued to slow the travelers across Wyoming. On Monday, June 5, a brief, furious snowstorm surprised Old Steady's crew 10 miles west of Elk Mountain. Four days later, Megargel and Stanchfield were caught in a hailstorm west of Granger. "Both of us

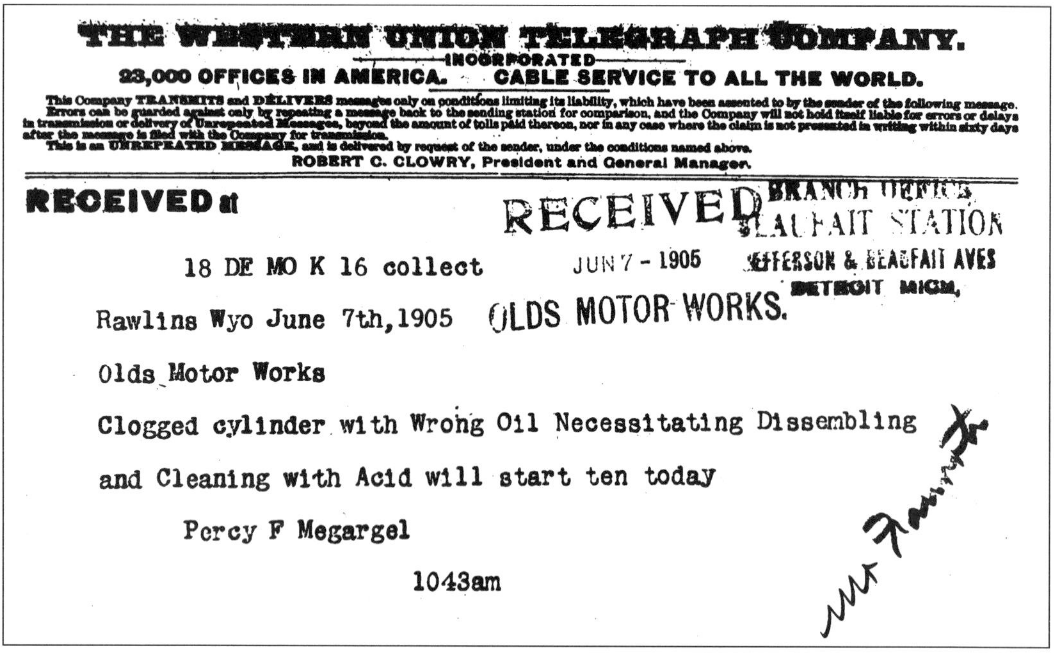
THE WESTERN UNION TELEGRAPH COMPANY.
INCORPORATED
23,000 OFFICES IN AMERICA. CABLE SERVICE TO ALL THE WORLD.
This Company TRANSMITS and DELIVERS messages only on conditions limiting its liability, which have been assented to by the sender of the following message. Errors can be guarded against only by repeating a message back to the sending station for comparison, and the Company will not hold itself liable for errors or delays in transmission or delivery of Unrepeated Messages, beyond the amount of tolls paid thereon, nor in any case where the claim is not presented in writing within sixty days after the message is filed with the Company for transmission.
This is an UNREPEATED MESSAGE, and is delivered by request of the sender, under the conditions named above.
ROBERT C. CLOWRY, President and General Manager.

RECEIVED at

RECEIVED BRANCH OFFICE, BEAUFAIT STATION
JUN 7 - 1905 JEFFERSON & BEAUFAIT AVES DETROIT MICH.
OLDS MOTOR WORKS.

18 DE MO K 16 collect

Rawlins Wyo June 7th,1905

Olds Motor Works

Clogged cylinder with Wrong Oil Necessitating Dissembling and Cleaning with Acid will start ten today

Percy F Megargel

1043am

Fig. 1.17 Megargel telegraphed news of Old Steady's troubles. The handwritten name in the lower right corner refers to Fay L. Faurote, Olds advertising manager. (OHC)

crawled under the car in the mud to save ourselves from the chunks of ice nearly two inches in diameter that descended in torrents," Megargel wrote. Two inches of ice littered the ground before the hail turned to a heavy rain. The men continued but soon stalled Old Steady in a raging mountain stream. "We half waded and half swam to shore, leaving the car with just the top of the seat above water. For some hours we waited for the storm to abate, but the longer we waited the higher the water rose."

In search of help, Megargel and Stanchfield hiked to a sheep ranch, whose owner Megargel calls Lone John:

> He seemed glad to see us, as company is scarce in this part of the world, and immediately commenced to tell us his troubles, which were numerous. He was at war with several of his neighbors, and kept a loaded gun at hand at all times. He had just been released for cutting open the head of one of his neighbors with an axe, and he regretted the fact that he had not killed the man. Despite his grievances, he willingly threw the harness on his horses, and telling us where we would find the wagon, and to use his team as we saw fit, went out to round up his flock of about 500 sheep and half that number of lambs.[93]

Old Scout's crew left Sage, Wyoming, on Tuesday, June 6 after spending the night at the house of the railroad section boss. (See Fig. 1.18.) "We took the main road along the railroad and drove about five miles, when the road left the railroad and started off through the sage brush," Huss wrote. "As we could see no other road we were sure we had the right one. We drove two hours and landed up against a large ranch down between some mountains. We asked the distance to Cokeville and found we were farther away than when we left Sage. We were both very much disgusted, as we had to go back over the same trail to the railroad. After skirmishing around in the sage brush for some time we found the tracks and arrived in Cokeville at noon, having covered 50 miles, and going only 15 in the right direction."[94]

Huss and Wigle crossed the Idaho border to reach Soda Springs at 7 p.m. Tuesday. Some of the larger towns on their route west included McCammon, Pocatello, Blackfoot, Arco, Hailey

Fig. 1.18 According to one source, "'Old Scout' excites the curiosity of a group of school teachers vacationing at a ranch near Granger, Wyoming," east of Sage. (OHC)

and Boise in Idaho; and Ontario, Burns, Prineville, Sisters and Salem in Oregon. The next day, Wednesday, June 7—the day Old Steady left Rawlins for the second time—Old Scout had a mishap near McCammon on a steep hill, "so steep in fact, that it was impossible to hold the car," Huss recalled. "We shot down the hill and over a 12-foot embankment into a ditch. Our good luck was that the ditch was dry. Wigle and I had a hard time getting the car out, and were surprised to find it still hanging together. The front axle was so badly bent that the wheels were on a slant of fifteen degrees."[95] The men coaxed the car the 30 miles into Pocatello and spent much of Wednesday night and Thursday morning repairing the damage.

Salvation by Sand Tires

Soon after passing Ross Forks, Idaho, on Thursday, June 8, Day 32, Huss and Wigle found themselves "hub deep in the sand, and Old Scout gave a few grunts and stopped. As we had carried sand tires, we thought this was a good chance to test their efficiency.... The way Old Scout performed with them was a wonder."[96] Sand wasn't the only obstacle, Huss revealed in Boise: "On the desert between Blackfoot and Arco we encountered a terrific storm. We sought shelter twice, being nearly frozen."[97] (See Fig. 1.19.)

Megargel and Stanchfield also resorted on occasion to sand tires, which Lester L. Whitman introduced to transcontinental driving during his 1903 Olds trip. In a 1908 fiction book based on his transcontinental experiences, Megargel describes sand tires: "Made of canvas stuffed with cotton-waste they strapped around the rims, the Goodrich pneumatics filling out the center of the sand-tires, thus giving to each wheel a tread of several additional inches."[98]

Arriving in Boise by train Saturday evening, June 10, Abbott told a Boise newspaper that Huss and Wigle had reached Corral, Idaho, where they evidently spent Saturday night. When the auto reached Boise at 4:15 p.m. Sunday, it looked true to its name of "Old" Scout, according to Monday's *Boise Evening Capital News*:

Fig. 1.19 Wigle, center, tends to Old Scout at a June 9 overnight stop in Martin, Idaho. The woman ran the Martin post office; the man at right is unidentified. (OHC)

The appearance of the machine is not what it was when it was taken out of the factory to start on the long journey. As it pulled into Boise yesterday it was covered with dust, the wheels were dished as the result of coming in contact with huge boulders on the way, and the tires looked more like a wornout mop than the article the factory turns out. The driver and his companion looked as if they had passed through a Nebraska sandstorm…. They declare they will be glad when the race is ended and that they do not care to undertake another.[99]

The *Capital News* and *Idaho Daily Statesman* of Boise both reported that Old Steady was in western Wyoming, far behind and "out of the race," though still planning to finish the drive to Portland.[100]

An Awful Hazard to Health

Newspaper accounts and the chronicles of Huss and Megargel indicate that all four men had health problems on the western half of the trip. Stanchfield caught "mountain fever" at an 8,000-foot elevation in western Wyoming—"aches in every bone and muscle, has a burning fever with intermittent cold chills, and was most miserable," Megargel recalled.[101] Megargel left Stanchfield at a hotel in Diamondville, Wyoming, on Saturday, June 10, and drove the car on what his *Automobile* diary said was a 20-mile round trip to Kemmerer for a doctor. Megargel continued alone until Monday, June 12, he says in *Automobile*, when Stanchfield arrived by train at Soda Springs, Idaho, to rejoin Old Steady. (See Fig. 1.20.)

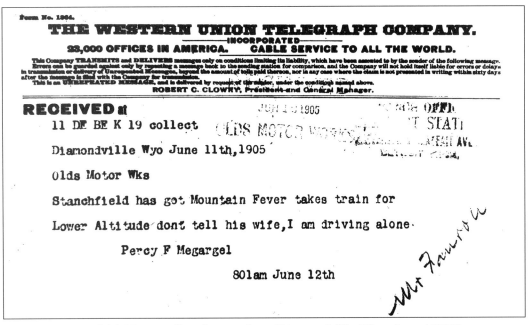

Fig. 1.20 Megargel's telegram from Diamondville, Wyoming. (OHC)

Megargel also had problems. "The different altitudes did not affect me until nearing Soda Springs, when, after pumping up one of our tires, I suffered from a violent bleeding of the nose," Megargel said. "The slightest exertion causes a person to pant violently, while harder work causes one not accustomed to the mountains to bleed at the nose, mouth and ears."[102]

The June 17 *New York Times* reported that "Huss left Boise City early in the week after taking a day's rest, as he has been suffering from chills, resulting from his long exposure to changeable weather." "My partner, Mr. Wigle, was sick many times from overexerting himself," according to Huss, who afterward told the *Detroit Evening News* of another incident: "In crossing some of the deserts in the west we were badly in need of water. For two days we had scarcely anything to drink, and when we walked we wabbled. It was my first experience of the kind, and I was dizzy for days afterwards."[103]

Huss doesn't mention this desert incident in the *Hell Gate to Portland* booklet, and also avoids mentioning his own health problems. *Motor World*, however, reported that their six-day lead allowed Huss and Wigle to rest in Boise on Monday, June 12. "Huss was loath to do so … and it was only at the earnest solicitation of friends that he consented to do anything of the sort. The plucky fellow is really in bad shape, and for the last four days has been going on his nerve alone."Caught in a cloudburst near Pocatello, Huss "came near having pneumonia," Abbott told the *Portland Evening Telegram*.[104]

The racers made "one serious mistake," Abbott contended. "In their zeal to make progress they traveled through storms for which they ought to have laid up. It is an awful hazard to health, vitality and machinery to try to plow through rain and mud as they did at times, and in my judgment the ultimate resultant of progress was loss."[105]

Battered, Patched and Wired Together

Old Scout resumed its journey from Boise at 1:45 p.m. Tuesday, June 13, expecting to reach Ontario, Oregon, before nightfall, the *Idaho Daily Statesman* reported. "A breakdown occurred, however, before Star [Idaho] was reached, and it became necessary to send back to Boise for repairs."[106] For the next five days, Old Scout traveled from western Idaho into west-central Oregon.

When Old Scout pulled into Prineville, Oregon, Sunday afternoon, June 18, it was "mud splattered, battered, worn, patched and wired together, but still capable of continuing on its race to the exposition city," observed Prineville's *Crook County Journal*. "'We've lost everything but the front wheels and our nerve,' said Chauffeur Huss while busily engaged in tightening up a few loose bolts.… Night runs were dispensed with some time ago as the head light on the machine was lost while traveling through sage brush that completely covered the auto and stood considerably higher than the machine itself."[107] (See Fig. 1.21.) Old Scout left after staying an hour in Prineville, the newspaper said, and spent Sunday night in Forest, Oregon, on the eastern slope of the Cascades.

To reach Portland through the mountains, Old Scout on Monday, June 19, followed the historic Santiam Wagon Road over Santiam Pass (4,817 feet). Many local newspaper accounts claimed Old Scout was the first auto to cross the Cascades, at least over the Santiam Wagon Road. It was on this route that occurred "by far the most exciting adventure of our trip,"[108] Huss contended.

To move cattle from the Willamette Valley to central Oregon's rich pastures and gold-mining camps, promoters in the 1860s began building the Santiam Wagon Road, known formally as

Fig. 1.21 A searchlight that bolted to Old Scout's left seat frame is missing in this Oregon photo, as is the tank of compressed acetylene gas that powered the light. To save weight after losing their light, Huss and Wigle evidently decided to discard the tank. (OHC)

the Willamette Valley and Cascade Mountain Road. In 1865, workers had completed the section between Albany and Sisters in central Oregon, though promoters pronounced the toll road open for travel as far east as the Idaho border, according to Tony Farque, a U.S. Forest Service archaeologist.[109] Named for the nearby South Santiam River, the Santiam Wagon Road was used by autos as late as 1939, when U.S. 20 opened in Oregon.

In his accounts, Huss said that the Santiam Wagon Road took Old Scout over Sand and Stone mountains—evidently meaning Sand Mountain on the east and Iron Mountain farther west, for Farque said there is no Stone Mountain in the area. One hazard of this route was the steep Seven Mile Hill, named not for its length but for the nearby Sevenmile Creek, according to Wayne Shilts, a U.S. Forest Service engineer. Wigle walked behind, carrying the car's leather tool bag, as Old Scout inched up the road's steep grades. "Just as the car was about to stall, Wigle would drop the bag under a rear wheel and lean his weight against the car to hold it. After the engine cooled somewhat I would speed it, throw in the clutch and climb again," Huss recalled in a September 1931 *Oregon Motorist* article. "Sometimes we made as little as 50 yards at a time in this manner."[110]

Over the Cascades

The descent was as fast as the ascent was slow, Huss related in his 1931 account:

> Eventually we reached the top of Stone [Iron] Mountain and thought our troubles were about over. Wigle and I cheerfully started driving down the slope. But we had not reckoned on the steepness of the grade. Our brakes were practically powerless and twice "Old Scout" was entirely out of control. I shouted to Wigle that we would have to jump if the grade did not ease up. We got down safely—just how I do not know....

> Sand Mountain was negotiated in the same manner [going up], only this time we eliminated the thrills of the descent in rather a novel fashion. When we reached the summit we cut down a good-sized fir tree and chained it to our rear axle. With Wigle riding this drag brake, we got down the mountain without difficulty or danger.[111]

Huss evidently rewrote history slightly in preparing his 1931 account. In a 1905 speech upon returning to Detroit from Portland, he said nothing about cutting down a tree: "Going down the other side of the Cascades we had a seven-mile hill so steep that I dragged Wigle at the end of a rope to keep us under check, though I had the brakes set all the time."[112] Similarly, in his 1905 *From Hell Gate to Portland* version, Huss said "Wigle, who was holding to the rear, was dragged nearly seven miles"[113]—and makes no mention of a felled tree. Using a tree brake was less novel than Huss suspected, regardless. During their 1903 Oldsmobile coast-to-coast trip, Lester L. Whitman and Eugene I. Hammond tied a pine tree to their Oldsmobile's rear axle in the Sierra Nevada. Teamsters had long used the same method to slow their wagons.

At about 9 p.m. Tuesday, June 20, the race leaders reached Salem, where they stayed overnight. "The dusty machine, with covered gear almost swimming in oil, was the center of much interest," according to a newspaper dispatch from Salem. (See Fig. 1.22.) "Souvenir hunters carefully picked out of the machine twigs of sage brush caught in the box and carried across the mountains from Eastern Oregon." Huss managed to work in a plug for the light Oldsmobiles, which he said had the advantage over heavier, more powerful autos on the roads Old Scout had traversed. "A large machine would not stand the jolting over rocks and ruts the way the light machines do."[114]

Fig. 1.22 Even the marshal in this unidentified western town dropped what he was doing when Wigle (pictured) and Huss appeared in Old Scout. (OHC)

Portland at Last

Old Scout left Salem at 7:30 a.m. Wednesday, June 21, and at Oregon City met a caravan of Portland Automobile Club members, who escorted the weary racers to Portland. Their 12:58 p.m. arrival in Portland[115] came 44 days, 6 hours, 28 minutes after leaving New York City (compensating for the three-hour time difference between the cities). This was well behind the record of just under 33 days that Lester L. Whitman and Clayton S. Carris had set in 1904.

The actual running time was 40 days, according to Huss. The record, however, shows that Old Scout's trip consumed 45 calendar days, three of which were spent idle—two days in Omaha and one in Boise. Old Scout's running time was thus 42 days. He and Wigle arrived 62 minutes before the start of the fifth annual National Good Roads Convention, held during Portland's Lewis and Clark Centennial Exposition. As part of the convention, experts would demonstrate new techniques by constructing a section of model macadam roadway on the expo grounds.

"In the midst of thousands of cheering spectators, Huss and Wigle of Detroit drove Old Scout through the streets of this city and into the grounds of the exposition," *Automobile* said in a Portland-datelined article. The transcontinentalists were "very much tanned and weatherbeaten, but in excellent health," according to the *Portland Morning Oregonian*.[116] "Our reception was wildly exciting, and the streets were packed with cheering people," Huss recalled. (See Fig. 1.23.) As a convention delegate from Michigan, Huss delivered to Henry W. Goode, president of the Lewis and Clark Exposition Company, a message from Associated Press Manager Melville E. Stone. (See Fig. 1.24.) Old Steady also carried a copy. The message read in part:

Fig. 1.23 Wigle, left, and Huss outside the Portland Olds agency. Their ribbons say "Portland Automobile Club/Good Roads Committee." (OHC)

The century which has just passed was chiefly notable as the century which developed inter-communication. It was the century out of which came the ocean steamship, the railway, the telegraph, the cable, the telephone, wireless telegraphy, rotary and perfecting printing presses, stereotyping, news gathering associations and the newspapers.

But nothing could better illustrate the progress of 100 years than a comparison of this new expedition by a twentieth century motor car from the Atlantic seaboard to the land "where rolls the Oregon," with that other expedition of Lewis and Clark, which meant so much for our common country and the world's civilization.[117]

Huss and Wigle were also persuaded to describe their experiences in Old Scout. As the *Portland Morning Oregonian* recorded a portion of Huss' remarks:

Fig. 1.24 Wearing a ribbon on his dark suitcoat, Henry W. Goode greets Huss, left, and Wigle in Portland on June 21, 1905. James W. Abbott, wearing a white coat, is partly obscured behind Goode. (FLP)

Mr. Huss was clad in a leather jacket, and gave every evidence of having passed through a very trying ordeal in making his record-breaking trip across the continent.

"You have better roads in Oregon than in any other state west of Chicago," began the automobilist, and the local delegates to the convention drowned out further remarks in a roar of applause. When the noise had subsided the speaker continued: "I am satisfied that with a little work, Oregon could have as good roads as New York and Ohio."[118]

Ironically, in another article on the same page, the *Oregonian* revealed that Huss and Wigle "report no part of the way more impassable than the Cascade Range road."[119] Huss said that after the speechmaking,

Although deadly tired, dirty and unshaven, it was impossible for Wigle or myself to escape the friendly captivity of the Portland enthusiasts. There was a monster parade, headed by "Old Scout" through the streets of Portland and the exhibition grounds.

Later we managed to evade our enthusiastic friends and escape to bed. Our battered shoes and hats and our soiled and stained clothing were virtually taken from us by storekeepers, who gave us complete new outfits for the privilege of exhibiting our apparel in their windows as the belongings of the first men to win a cross-country race in an automobile. We stayed in Portland about three days and then entrained, homeward bound.[120]

"You would be astonished at what an automobile will stand and still be in working order," Huss told *Motor World*. "Once we came to a stream where there was no bridge and the water was four feet deep. It seemed useless to try to ford it, but we took a swift run and plunged in. The momentum carried the automobile through the stream. We poured the water out of the engine and were going again in fifteen minutes."[121] Back home again, Huss told the *Detroit Evening News* about rounding a curve in Oregon's Cascade Range and finding an "immense rock" in the path ahead:

> To the right was a sheer wall running up hundreds of feet, while on the left was a ravine 300 feet deep. We couldn't go over it, so took a chance and tried to get around on the open side. There happened to be some sand there, we slipped, and our left wheels slid off and spun around in space. If we had slipped a few inches more we would have never finished that race. We climbed out, chipped the rock away, set our block and tackle and in an hour were on our way again.[122]

Old Steady Arrives

Megargel and Stanchfield began their climb to the Cascade summit at Santiam Pass on Monday, June 26, intending to take the advice of the locals who suggested they drag a tree to slow their descent of Seven Mile Hill. Lacking an ax, they used a monkey wrench and chisel to fell a tree. But they cut it too early, and used it on what turned out to be three short downhill stretches. After the third, they found "a still steeper and higher hill to climb, so we cast our tree aside and went the rest of the way, including the seven-mile stretch, with only brakes and reverse. Everything went well until we were descending the sixth mountain range, when Old Steady commenced to skid" on a wet 50-percent grade, Megargel said.

> Down, down we came, first the front wheels ahead, and then the rear. Finally the front wheel struck a moss-covered rock by the roadside and climbed over. The other front wheel, not to be left behind, followed suit, as did one rear wheel. When we first struck, both Stanchfield and myself were not only thrown out, but pitched bodily clear across the road into the bushes, followed by most of our trappings.
>
> The car, suspended by one rear wheel and the right step, which had caught in some bushes, hung over a 200-foot cliff, and Stanchfield and myself, rubbing the mud out of our eyes, sat on the side of the road, wondering why it did not go all the way over.[123]

The occupants of a passing prairie schooner helped the men save the car; it was raining hard the entire time.

Portland's old-clothes craze was still strong when Old Steady's crew arrived on Wednesday, June 28. "We were recognized in the shoe store we patronized," Megargel said, "and nothing would do but for us to exchange our old shoes for new, the old ones going in the window for exhibition."[124] With Old Steady's finish, Oldsmobile became the first make of auto to cross the continent more than once—not just two but three times, beginning with the Whitman-Hammond 1903 San Francisco–New York journey in an Olds runabout.

Despite its tardiness, Old Steady received a royal welcome as it drove the last miles to Portland, arriving "shortly after 5 o'clock"[125] in the afternoon on June 28. "Leaving Sweet Home, we had traveled but little more than a mile when we were surprised by a young lady, who handed a bouquet of roses to Stanchfield.… It brightened Stanchfield's muddy face wonderfully," Megargel related. "From that time until Salem was reached flowers were given us every few miles, and Old Steady took on holiday attire."[126]

In late 1904, Lester L. Whitman and C.S. Carris in a Franklin car had set the standing transcontinental record of 32 days, 17 hours, 20 minutes. Megargel had hoped to cross in one month but June 8, his 31st day of elapsed time, found him in Granger, Wyoming, 1,000 miles short of his goal. "Had the weather been at all favorable, I still think we could have made the run in one month,"[127] Megargel contended at the time.

Only after arriving in Prineville, Oregon, on Saturday, June 24, did Megargel and Stanchfield learn that Old Scout had finished the race three days earlier, the *Portland Morning Oregonian* reported. So "from Prineville to Portland Megargle [sic] and his companion slowed up and took their time to reach Portland."[128] Old Steady's elapsed time to Portland was 51 days and about 10½ hours (compensating for time changes en route). The trip thus consumed 52 calendar days. Old Steady was idle for four of those days—two in Omaha, one in Cheyenne and one in Rawlins—giving it a running time of 48 days. (See Table 1.1.) Because the runabouts lacked odometers, the exact mileage traveled in the first transcontinental auto race is unknown. Press accounts, however, estimated that the cars traveled between 3,500 and 4,000 miles. Assuming a compromise figure of 3,750 miles, Old Scout thus averaged 3.53 mph and nearly 85 miles per day for its elapsed time, while Old Steady averaged 3.04 mph and nearly 73 miles per day.

Bad Roads Deliver Hard Knocks

"Old Scout is, to all appearances, none the worse for the hard knocks it has received," *Automobile* asserted, "and on the run into the city covered one stretch of twenty-five miles in one hour flat." The Olds factory in later years would repeat this fiction—claiming that both cars arrived "none the worse for wear"—but knew better, as an owner's manual for early 1-cylinder Oldsmobiles reveals: "Always pick out good roads for your machine; to run over stones and rough places means unnecessary wear and tear."[129]

Both runabouts had just endured several thousand miles of such unnecessary wear and tear. Megargel saw the winning Olds runabout when he reached Portland: "It plainly showed the hard knocks it had received en route,"[130] he observed.

It was clear, however, that America's neglected or virtually nonexistent roads had delivered the hard knocks that Old Scout and Old Steady had absorbed. "One who follows carefully the story of this race, will get a fairly accurate idea of the road conditions of the country, and will be impressed with the need of a concerted good road movement," Olds Motor

Table 1.1 Daily Progress of the 1905 Oldsmobile Racers

Day/Date		Old Scout	Old Steady
1	May 8 Mon	New York City–Albany, N.Y.	New York City–Albany, N.Y.
2	May 9 Tue	Albany–Syracuse, N.Y.	Albany–Syracuse, N.Y.
3	May 10 Wed	Syracuse–Buffalo, N.Y.	Syracuse–Buffalo, N.Y.
4	May 11 Thu	Buffalo, N.Y.–Painesville, Ohio	Buffalo, N.Y.–Conneaut, Ohio
5	May 12 Fri	Painesville–Clyde, Ohio	Conneaut–Clyde, Ohio
6	May 13 Sat	Clyde, Ohio–Kendallville, Ind.	Clyde, Ohio–Kendallville, Ind.
7	May 14 Sun	Kendallville–Hammond, Ind.	Kendallville–La Porte, Ind.
8	May 15 Mon	Hammond, Ind.–Aurora, Ill.	La Porte, Ind.–Aurora, Ill.
9	May 16 Tue	Aurora–Mendota, Ill.	Aurora–Mendota, Ill.
10	May 17 Wed	Mendota–Geneseo, Ill.	Mendota–Geneseo, Ill.
11	May 18 Thu	Geneseo, Ill.–Wilton, Iowa	Geneseo, Ill.–Wilton, Iowa
12	May 19 Fri	Wilton–Brooklyn, Iowa	Wilton–Brooklyn, Iowa
13	May 20 Sat	Brooklyn–Des Moines, Iowa	Brooklyn–Des Moines, Iowa
14	May 21 Sun	Des Moines–Dexter, Iowa	Des Moines–Dexter, Iowa
15	May 22 Mon	Dexter, Iowa–Omaha, Neb.	Dexter, Iowa–Omaha, Neb.
16	May 23 Tue	**Omaha: repair/prep**	**Omaha: repair/prep**
17	May 24 Wed	**Omaha: repair/prep/rain**	**Omaha: repair/prep/rain**
18	May 25 Thu	Omaha–Fremont, Neb.	Omaha–Fremont, Neb.
19	May 26 Fri	Fremont–Grand Island, Neb.	Fremont–Palmer, Neb.
20	May 27 Sat	Grand Island–Kearney, Neb.	Palmer–Grand Island, Neb.
21	May 28 Sun	Kearney–Cozad, Neb.	Grand Island–Kearney, Neb.
22	May 29 Mon	Cozad–North Platte, Neb.	Kearney–Cozad, Neb.
23	May 30 Tue	N. Platte, Neb.–Julesburg, Colo.	Cozad–Sutherland, Neb.
24	May 31 Wed	Julesburg, Colo.–Cheyenne, Wyo.	Sutherland–Lodgepole, Neb.
25	June 1 Thu	Cheyenne–Laramie, Wyo.	Lodgepole, Neb.–Cheyenne, Wyo.
26	June 2 Fri	Laramie–Medicine Bow, Wyo.	**Cheyenne: repairs**
27	June 3 Sat	Medicine Bow–Creston, Wyo.	Cheyenne–Terry Fee Ranch[1]
28	June 4 Sun	Creston–Green River, Wyo.	Terry Fee Ranch–Elk Mountain, Wyo.
29	June 5 Mon	Green River–Sage, Wyo.	Elk Mountain–Rawlins, Wyo.
30	June 6 Tue	Sage, Wyo.–Soda Springs, Idaho	**Rawlins: repairs**
31	June 7 Wed	Soda Springs–Pocatello, Idaho	Rawlins–Point of Rocks, Wyo.
32	June 8 Thu	Pocatello–Blackfoot, Idaho	Point of Rocks–Granger, Wyo.
33	June 9 Fri	Blackfoot–Martin, Idaho	Granger– ?
34	June 10 Sat	Martin–Corral, Idaho	? –Diamondville, Wyo.
35	June 11 Sun	Corral–Boise, Idaho	Diamondville–Cokeville, Wyo.
36	June 12 Mon	**Boise: rest day**	Cokeville, Wyo.–Soda Springs, Idaho
37	June 13 Tue	Boise–Star, Idaho[2]	Soda Springs–McCammon, Idaho
38	June 14 Wed	Star–Nampa, Idaho	McCammon–Pocatello, Idaho
39	June 15 Thu	Nampa, Idaho–Ontario, Ore.	Pocatello–Blackfoot, Idaho
40	June 16 Fri	Ontario–Drewsey, Ore.	Blackfoot–Cottonwood Ranch, Idaho[3]
41	June 17 Sat	Drewsey–East of Paulina, Ore.[4]	Cottonwood Ranch– ?
42	June 18 Sun	East of Paulina, Ore.–Forest, Ore.	? – ?
43	June 19 Mon	Forest–Mountain House, Ore.[5]	? –Boise, Idaho
44	June 20 Tue	Mountain House–Salem, Ore.	Boise, Idaho–Ontario, Ore.
45	June 21 Wed	Salem–<u>Portland, Ore.</u>	Ontario–Johnson's Ranch, Ore.[6]
46	June 22 Thu		Johnson's Ranch–Buchanan Ranch, Ore.[7]
47	June 23 Fri		Buchanan Ranch–Stage Station[8]
48	June 24 Sat		Stage Station–Prineville, Ore.
49	June 25 Sun		Prineville–Cache Creek, Ore.[9]
50	June 26 Mon		Cache Creek–Sweet Home, Ore.
51	June 27 Tue		Sweet Home–Salem, Ore.
52	June 28 Wed		Salem–<u>Portland, Ore.</u>

[1] The Terry Fee Ranch was west of Laramie, Wyo.

[2] Old Scout, which left Boise at 1:45 p.m. June 13, broke down before reaching Star, Idaho, "and it became necessary to send back to Boise for repairs," said Boise's *Idaho Daily Statesman* of June 14. The newspaper did not say if Old Scout spent the night in Star.

[3] Megargel also identified Cottonwood Ranch, situated west of Arco, Idaho, as Martin Post Office—evidently Martin, Idaho.

[4] Old Scout spent the night 15 miles east of Paulina, or about midway between Suplee and Paulina, according to the June 22 *Crook County Journal* of Prineville, where Old Scout arrived Sunday.

[5] Old Scout spent most of the day Monday, June 19, climbing the Cascades to the summit. Huss and Wigle stopped for the night at "the mountain house seven miles down the west slope of the mountains," said the June 21 *Daily Oregon Statesman* of Salem, where Old Scout stopped Tuesday night, June 20.

[6] Johnson's Ranch was about 10 miles west of what Megargel called Vale Post Office, Ore.

[7] George Buchanan's Ranch in Harney County, east of Harney, Ore.—most likely in or near the town of Buchanan, shown on some modern highway maps as situated a few miles east of Harney.

[8] This Crook County stage station was either 68 or 84 miles east of Prineville, according to conflicting information Megargel gave in his July 13 *Automobile* account.

[9] Cache Creek was the eastern toll station for a toll road that Old Steady followed—free of charge—over the Cascades.

Sources: Megargel's *Automobile* accounts; local newspapers; telegrams sent from the racers to Olds Motor Works; and Huss and Megargel in the factory booklet, *From Hell Gate to Portland*.

Works concluded.[131] Huss was more direct: "My remarks about some of the roads over which we passed would be unfit to print."[132]

In his *Automobile* diary, Megargel had little to say about routine repairs; the telegrams the autoists sent from the road to Olds Motor Works were more revealing. "Express me Omaha large gear pump one on car no good," Huss wired from Chicago. "Must have new tires new steering spring and good new carbureter at Omaha ship immediately," Megargel telegraphed from Brooklyn, Iowa.

Brooklyn is where Huss and Wigle repaired a broken drive chain, "which is the first breakage they have had on the trip of over 1,300 miles so far,"[133] according to *Motor World*. "Our machine has met only two accidents," Huss said in Salem, "one the breaking of a spring, which caused a delay of less than two hours, and the other the breaking of [our] gear, causing a delay of a day and a half…. Our repairs have cost us about $15,"[134] though he was quoted elsewhere as saying $17. In interviews after the trip, Megargel said Old Steady used three sets of Diamond tires and Huss said Old Scout used two sets of Fisk tires.

When Old Steady reached Prineville on Saturday, June 24, to spend the night, a *Crook County Journal* reporter asked about the cost of the race. "Mr. Megargel said that the affair had cost the Oldsmobile company in the neighborhood of $14,000, which is all applied to the advertising account. Mr. Megargel and his assistant are each receiving $300 per month and expenses for driving the machine. At Burns gasoline cost them 80 cents a gallon, and their expense has run into three figured numbers every week during the month and a half since they left New York City."[135]

Later Exploits

Both Huss and Megargel distinguished themselves in later driving contests. Huss helped drive an Oldsmobile Model S in a 200-hour nonstop test in Detroit the year after his transcontinental victory.[136]

During the 1907 Glidden Tour, the American Automobile Association's annual endurance contest, a Packard overturned on a sharp turn near Bryan, Ohio, fatally injuring its driver. Huss stepped in to drive the damaged car, *Automobile* recounted:

> Half of the steering wheel had been carried away in the fall, making the car exceptionally difficult to manage on the greasy [muddy] roads. At Brimfield [Indiana] there is a stiff, short rise, followed by a sharp turn to a wooden bridge crossing the railroad. Although going at slow speed, it was impossible to control the Packard, which crashed into the right-hand side of the bridge, breaking down the wooden barrier and stopping with the front wheels suspended over a twenty-foot drop to the railroad below.... [T]he machine was hauled back onto the road, repaired, and once more started for South Bend.[137]

At the September 1908 Toledo–Columbus–Cleveland three-day reliability run in Ohio, Huss was the "surprise of the run," turning in a perfect score driving a Brush runabout. "This is considered remarkable, for when it is remembered that this car carried only a 7-horsepower single-cylinder engine the fact that it climbed the hills and worked through the awful sand seems incredible,"[138] *Motor Age* raved. Of course, Huss got plenty of practice during his 1905 trip in the 7-horsepower Olds.

From 1913 until his retirement in 1930, Huss worked for the makers of the Hupmobile, the Hupp Motor Car Company, where he became production and sales engineer, according to one source.[139] With Oldsmobile's backing in 1931, Huss repeated his transcontinental drive in Old Scout, following approximately the same route as in 1905. Huss was escorted by three 1931 Oldsmobile sedans at a leisurely pace to allow for frequent stops at Olds dealerships and for receptions. Oldsmobile advertised it as a Transcontinental Good Roads Tour "for the purpose of encouraging further the good roads movement throughout the nation."[140]

In 1937, the 64-year-old Huss, who gave his age as 49, took a job at Ford. He worked the next 15 years at Ford's River Rouge plant, where he eventually took charge of all piston-grinding operations, according to his son, John. Learning his true age through Social Security records, Ford officials stopped him in the middle of work one day, informing him that he'd have to retire immediately—but then kept him on the payroll another six months while he trained his replacement.[141] Huss retired from Ford in 1952, according to a *Life* magazine feature. He died in Detroit at age 90 on August 19, 1964.[142]

In late 1905 and early 1906, Percy Megargel and a partner drove a Reo in the first winter coast-to-coast trip and the first double transcontinental crossing. On May 2, 1909, at 34 years of age, Megargel died of cancer, leaving behind a wife and infant daughter. He was buried at Scranton, Pennsylvania, his boyhood home.[143]

Fate of the Cars

After the race, Olds Motor Works sold the winning Old Scout to E. Henry Wemme, a Portland tent and awning manufacturer who was "a pioneer automobile enthusiast and holds the distinction of having owned the first machine in Oregon," *Automobile* said. "It is very appropriate that he should now own the first automobile that ever reached Oregon overland from New York."[144]

For placing second in the 1905 race, Megargel kept Old Steady. But a retrospective article in *Antique Automobile* contends that Wemme "also purchased the second Olds, Old Steady. He and an employee, Charles Wintermute, used to drive Old Scout and Old Steady in Portland. One day while Wintermute was driving Old Steady along Fifth Avenue, a horse suddenly stopped in front of him. Wintermute applied the brakes, the car spun around and was hit by a street car that had been following the helpless Oldsmobile. The auto was completely wrecked."[145]

A photo in the January 1923 *Oldsmobile Pacemaker* shows Old Scout powering a sawmill. "Old Scout is still running and when not occupied otherwise, it earns its 'board and keep' by sawing wood,"[146] boasted the factory, without naming the owner of the car. Shortly afterward, two Olds Motor Works officials who visited Portland posed for photos with Old Scout and Ed Cohen, identified on the back of a surviving photo as "head of the Oldsmobile Company of Oregon." In the photo, Old Scout carries a 1924 dealer plate. (See Fig. 1.25.)

But the Oldsmobile company had evidently lost track of the car by 1931. In trying to locate Old Scout so Huss could drive it in the 1931 reprise of the original trip, "a telegram was sent to E.E. Cohen, Portland distributor of Oldsmobile cars, asking if he knew what had happened to the ancient rig," according to the *Portland Sunday Oregonian*. "Cohen … replied, 'Old Scout is on my floor. Why do you want to know?'" Cohen had bought Old Scout from Wemme "because of its historical interest and from time to time has run it through Portland as an advertising stunt," according to the newspaper.[147] This was the car Huss used during his 1931 trip, according to the *Sunday Oregonian* and the 1931 *Oregon Motorist*.[148]

Roob H. Allie, who managed the 1931 reenactment, said Huss drove Old Scout the entire distance. Later, Oldsmobile arranged to have Old Scout driven from Lansing to Chicago's Century of Progress International Exposition in 1933. "Also it was on view at [the] Galveston World's Fair and the New York World's Fair and at several auto shows," Allie told the *Detroit Free Press*.[149] Old Scout was also among items exhibited on the American Freedom Train, which in 1975 began a two-year tour of the United States in honor of the nation's bicentennial.[150]

Oldsmobile owns a 1904 curved-dash Olds restored as Old Scout and on display at the R.E. Olds Transportation Museum in Lansing, Michigan. Because no record survives of Old Scout's serial number, officials cannot be absolutely certain of the restored car's heritage, according to Helen Earley of the Oldsmobile History Center in Lansing. But "there's some fairly good evidence that this is the car," according to Gary Hoonsbeen, an authority on curved-dash Oldsmobiles.[151]

Fig. 1.25 The back of this photo identifies two of these men—perhaps those on the far left—as Thomas O'Brien and Roy M. Hatfield "of the Olds Motor Works," who were touring the West. Ed Cohen is apparently one of the men in the car. The setting is Portland's Washington Park. (FLP)

Commemorating the Race

Part of the original route that Huss and Megargel followed over Oregon's Santiam Pass has survived the ravages of time and modernization. From 1993 to 1995, volunteers and U.S. Forest Service employees built five new bridges while restoring a 19½-mile stretch of the Santiam Wagon Road east from Sweet Home to Santiam Pass, said Tony Farque, a U.S. Forest Service archaeologist. Except for rerouting where U.S. 20 overlaps the original trail, the restored wagon road looks much as it did when Old Scout and Old Steady traveled it in 1905. Santiam Wagon Road is now open to horses, cyclists and hikers; drivers of wagons and pre-1940 autos may use the road by getting a permit from the Forest Service office at Sweet Home.[152]

With Dwight Huss' son, John, and Oldsmobile General Manager John D. Rock looking on, officials on July 7, 1995, dedicated an historical marker along the restored wagon road. A section of the trail was named Huss Ridge. Inspired by Edd Whitaker of Portland—who, while filming a video documentary titled *From Hellgate to Portland*, located John Huss in southern California—the historical marker is situated on the south side of U.S. 20 at the Lost Prairie Campground, about 38 miles east of Sweet Home.

The Race and its Reverberations

What did it all mean, a grueling race between two automobiles from New York City, that eastern center of culture and commerce, to Portland, the largest city in a mountainous, heavily forested state a continent away?

Six weeks after Old Steady reached Portland, James W. Abbott put the event he organized into historical perspective:

> There is going on to-day a movement toward the Pacific, involving more people than the gold excitement allured, even when the frenzy was at its height, and destined to be equally as important in epoch-making history. I believe the time to be close at hand when the automobile is to become a factor in this movement....[153]

Were he alive today, Abbott the prognosticator, who advocated spraying mineral oil on roadways to lay the dust and repel moisture, would be amazed at the changes wrought by two generations of automobile use. By 1960, it was possible to get almost anywhere without leaving a paved, two-lane road. Today, from New York City through much of Idaho, cars and trucks roar down a four-lane interstate highway that deviates only slightly from the route that Old Scout and Old Steady followed in 1905. In places, this concrete corridor intersects the path that Meriwether Lewis and William Clark followed with their band of explorers 100 years earlier. Driving nonstop on such roads, modern automobiles can travel farther in 24 hours than Old Steady and Old Scout traveled in two weeks or more.

When he reached Portland to win the 1905 race, Huss boasted that Old Scout had made the trip entirely under its own power. This is false. The May 18, 1905, *Chicago Evening Post* quoted Huss as saying, "In one place we had to hire a team to pull us out."[154] In his own writings, Huss documents many instances in which the men used a block and tackle or pushed and pulled the car to free it from mud, sand or water. Old Scout and Old Steady became the fifth and sixth automobiles to cross the country; none had done so entirely under its own power.

Besides demonstrating for the fifth and sixth times how bad U.S. roads were, the four drivers showed how ingenuity and resourcefulness could conquer seemingly impossible odds. The drivers also demonstrated that American cars could endure the roughest possible treatment. What's more, if Oldsmobiles were any indication, American cars were improving. During their transcontinental trip two years earlier, Hammond and Whitman advised R.E. Olds to improve the Oldsmobile's piston pins and rings, crankshaft and brakes. Olds apparently listened. Neither Old Scout nor Old Steady reported problems with these mechanical vitals.

The 1905 overland trek to the Lewis and Clark Centennial Exposition was both a race *and* a publicity stunt. It was a closed contest—open only to Oldsmobiles, in which the cars traveled together as far as eastern Nebraska. Muddy roads and identical power prevented one car from outdistancing the other. But who can doubt that each team was trying to wrest the $1,000 prize away from the other? If Old Steady had not broken through a bridge in central Nebraska, the race could have been much closer—or the outcome different. Regardless, for a mere $14,000, the retail price of 20 little runabouts, Oldsmobile bought its way into the Northwestern markets and spread its name and fame nationwide.

Being closed to all other automakers clearly limited the value of the race, both to the Olds Motor Works and to the automobile industry in general. Imagine the outcome if Huss had been able to actually prove his claim that large cars "would not stand the jolting over rocks and ruts the way the light machines do." Overnight, the curved-dash Olds could have become wildly popular. Observers of the 1909 New York-Seattle race would have *expected* the lightweight Ford Model T to beat its heavier rivals, and would have been surprised if it had not. For that matter, the Olds factory might have sped up production to flood the market with cheap, reliable curved-dash runabouts. Henry Ford, seeing no opportunity to compete in the low-price field, would have continued turning out relatively small numbers of heavy, expensive autos. There may have been no Model T....

Megargel's personal goal was to lower the transcontinental speed record. Rain and muddy roads prevented it. Nevertheless, sales prove that the effects of the 1905 race reverberated throughout the U.S. auto industry—and not just in the Northwest. Huss observed that Midwestern farmers flocked to towns on the route to watch the muddy cars drive through. Farmers were, in effect, shopping for an improved form of transportation. Thanks in part to the widely reported 1905 race, they found it. Nebraska auto-registration figures for 1905 to 1907 show that Oldsmobile topped them all, followed in order by Cadillac, Rambler, Reo (a later product of Ransom E. Olds, creator of the Oldsmobile), Ford and Maxwell.[155]

Three years passed before Americans witnessed another transcontinental contest: the 1908 New York–Paris race. Six cars from four nations entered that punishing event, which many historians still regard as the greatest automobile race of all time.

CHAPTER 2

An Act of Splendid Folly:
New York to Paris by Auto

This is more than a race; it is an expedition.

—Hans Hansen, contestant

Late in 1907, a Paris newspaper announced a 20,000-mile, nearly round-the-world auto race that included crossing the Bering Strait between Alaska and Siberia in winter "without taking the steamboat." (See Fig. 2.1.) The fanciful undertaking promised misery for men and machine alike. Nevertheless, six autos left New York City on February 12, 1908, to begin an overland expedition to Paris.

Contestants "must be men who are not easily overcome by difficulties, who are strong mentally and physically; able to use an axe to good purpose; who can wade through water up to their armpits, if necessary, in seeking to pull a machine out of a 'bad place,' who can withstand cold and possibly hunger without taking it too much to heart … and who, in conclusion, can adapt themselves to circumstances,"[1] advised the *New York Times*, which joined *Le Matin* of Paris in sponsoring the race.

Le Matin, which had sponsored a 1907 Peking-Paris race, "must be congratulated on having proved the impossible possible and for displaying more faith in the automobile than the most enthusiastic automobilist,"[2] *Automobile* magazine said. As did Charles Lindbergh's Paris-bound solo flight across the Atlantic in 1927, the 1908 race pitted man against nature in a fierce struggle that captivated the world.

It was a cheeky, breath-taking, brilliant newspaper stunt—and, like the Space Race of the 1960s, a very public scientific experiment. Just how much punishment could an automobile absorb? Like the Winter Olympics of later years, the 1908 Great Race pitted Americans, French, Germans and Italians against one another for the grand prize of glory. Eminent scientists, explorers, professors and automakers offered conflicting views of the possibilities. It was a delightful free-for-all that captured the world's imagination and generated nine months' worth of copy besides.

Many were quick to grasp the implications of the international contest, which also became the first U.S. transcontinental auto race involving more than one make. Automakers saw a chance

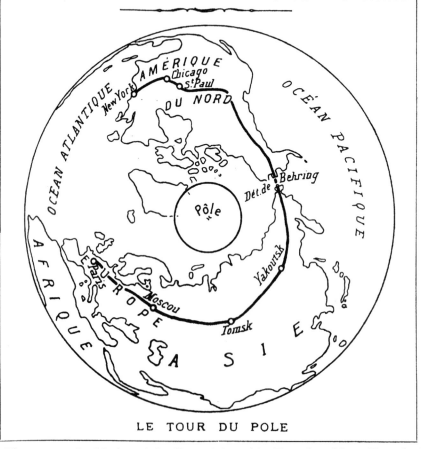

Fig. 2.1 The route as Le Matin *originally envisioned it. Translated from French: "The world circuit/New York–Paris by automobile without taking the steamboat/An automobile exploration around the world/Going around the pole."* (Im Auto um die Welt)

to prove the auto's usefulness. Scientists saw the contestants, traversing nearly unknown regions, as "pioneers in a vast field of scientific research."[3] What's more, a race across North America, Siberia and northern Asia to Europe, if completed, "would have an immense effect upon commercial relations between the two continents,"[4] the *New York Times* predicted. Alert schoolteachers used newspaper accounts as a daily geography primer. At least one participant, however, saw the race as a chance "to perpetrate an act of splendid folly, not to open up a new way for men. We wished to be madmen, not pioneers."[5]

Plan "Bristles with Difficulties"

Race organizers unintentionally kept the story lively by toying with many routes. Originally, the plan was to send the racers from New York to Chicago, and then northward to speed along 2,200 miles of the frozen Yukon River through Canada to Cape Prince of Wales, Alaska. There, they would cross an approximately 55-mile stretch of the Bering Strait on ice to reach East Cape, Siberia. Just one problem with this, volunteered an Alaskan mining engineer, an explorer and others: the Bering Strait rarely freezes solid—though it did back in 1884.

Others warned of ice hillocks on the frozen Yukon. Various experts suggested using Siberian ponies to pull the cars through Canada, or possibly disassembling the cars and walking the pieces over impassable terrain. A Siberian explorer advised packing tobacco and tea for northern Siberian natives, both to build goodwill and to barter for supplies. Racers would travel weeks between telegraph stations in Siberia, where the food "must give way to blubber—either that of whale, walrus or seal, as they may prefer, with occasionally a piece of reindeer meat," the *New York Times* advised. As one observer succinctly declared, "The plan bristles with difficulties."[6]

Finally, after two months of juggling alternatives, the *New York Times* suggested a more likely route through North America: From New York City westward to Reno, Nevada, along the route Lester L. Whitman used in 1906 to set the standing transcontinental record of 15 days, 2 hours, 12 minutes. At Reno, crews would decide whether to try crossing the snowy Sierra Nevada or detour some 900 miles south to Los Angeles on a surer route to San Francisco. From there, contestants would take steamer ships to Seattle and Valdez, Alaska, and travel along dogsled trails to Fairbanks, from whence the cars would travel partly on Yukon River ice to reach the Bering Sea at Nome. There, they could either disassemble their cars and pack them across the Bering Strait on dogsleds or wait till the first steamer of the season reached Nome.

Racers Allowed One Year

In crossing America—the only portion of the race considered in this chapter—contestants were expected to obey speed limits between New York City and Chicago. The

Times and American Automobile Association arranged for interpreters for foreign crews and pilots to guide all the cars. Extra baggage, repair parts and mail were forwarded to automobile clubs in cities on the route, or, if the crews ordered it, directly to the West Coast. To avoid cheating, racers would present special photo identification cards at 67 telegraph stations across the United States. Otherwise, the formal race rules did little besides stipulate "the crossing of the United States without any auxiliary power whatsoever unless during the passage through the Rocky Mountains and the Sierras."[7]

Raising the possibility that no one would reach Paris within the one year allotted, the rules stipulated that the car reaching "the point in the itinerary which is farthest from the starting point (New York) shall be declared the winner."[8] The foreign contestants—three of whom took part in *Le Matin's* 1907 Peking–Paris race—feared the snow, ice, frigid temperatures and desolation of Alaska and Siberia. American observers, however, saw as much danger in crossing the United States as winter melted into spring.

Percy Megargel and David Fassett were the only drivers who had made a wintertime U.S. transcontinental trip. On the eastbound leg of their 1905–06 trek in a Reo, the two men detoured 1,000 miles to avoid up to 30 feet of snow in the Sierra Nevada but still spent many days snowbound in the Arizona mountains. "If the cars agree to keep together across America, abandoning any car that hopelessly breaks down, I believe all of us can reach Siberia in safety, where the racing in earnest will begin,"[9] said Megargel, who tried in vain to get the backing of an automaker. Like Megargel, Whitman had made three transcontinental trips. Whitman's participation in the Great Race, however, was limited to helping plan the U.S. route.

Editorially, the *New York Times* tried to shake the foreign racers' obsession with Siberia: "There was no part of the Peking to Paris contest which covered mountains as precipitous as those which have to be crossed in the United States ... nor any stretch in which stress of weather conditions were so severe as the Winter weather the trip to 'Frisco is apt to meet."[10] The newspaper was all too correct, as two Peking-Paris racers dropped out of the round-the-world trek in America. And all the New York–Paris racers hired snow-shovelers and horses to help them cross the United States.

White Bear Oil? But of Course!

How to prepare and equip these automobiles and their crews? A *New York Times* "special agent" traveled to Alaska to explore the trail from Valdez to Fairbanks, and interview freighters, miners, mail carriers and others. His recommendation: "Every piece of unnecessary ornamentation should be discarded and the machine stripped down to a state where everything is sacrificed to usefulness and the roughest kind of usage."[11] The crew of the Thomas car heeded this advice from the start. Foreign teams eventually lightened their overloaded cars during the American leg of the journey.

Siberian explorer Captain Hans Hendrik Hansen, who entered the Great Race aboard the French De Dion car, suggested a compass and sextant for each auto to "avoid any possibility of any one going astray,"[12] as the *New York Times* paraphrased him. *Le Matin* favored equipping each car with a Winchester rifle, picks and shovels, spiked steel rims for ice driving and an under-chassis wicker platform to keep the car from sinking into snowdrifts. "The wheels, in case of soft snow being met with, could be supplied with wide wooden rims of water-wheel formation,"[13] the Paris newspaper proposed. It also suggested each man pack nine pairs of wool socks—and wear three at a time.

The *New York Times'* Alaska correspondent urged drivers to wear waterproof Eskimo "muckluck" boots, made from walrus and seal. (See Figs. 2.2 and 2.3.) In addition to wearing masks, goggles and reindeer-skin coats, some European entrants "intend to keep their own bodies coated with white bear oil throughout the arctic portion of the journey,"[14] wrote the *Times'* Paris correspondent.

Scramble to Enter (and Exit)

A flood of foreign entries greeted *Le Matin's* announcement of the race in late 1907. Three French automakers were among the earliest foreign companies to enter or declare their intention of doing so—Charron, De Dion and Sizaire-Naudin. In the ensuing weeks the *New York Times* named many more foreign autos as entrants or possible entrants, including the Itala that won the 1907 Peking–Paris race. Others were Breguet, Brouhot, Gobron-Brillie, Motobloc, Renault and Werner of France; Benz and Protos of Germany; and Brixia-Zust and Fiat of Italy.

The Hol-Tan car, made by the Moon Motor Car Company of St. Louis for sale to Easterners through the Hol-Tan Company of New York City, became the first American entry on November 27, 1907. This was two days after the *New York Times* carried a one-paragraph reference to the just-announced race. By mid-January, three more U.S. automakers—Maxwell, Thomas and White—had entered the race. Because Corbin, Franklin, Reo and Studebaker showed a belated interest, observed the *New York Times* on January 20, 1908, "There is every reason to believe that when the race starts the American competitors will outnumber the foreign."[15]

But Hol-Tan, Maxwell and White all withdrew during the same week, the February 6 newspaper said. Other interested American automakers failed to follow through. "There were many to volunteer entries when they thought the contest destined to be run on paper merely," the *Times* noted wryly, "but when it came to a serious trial they withdrew with dispatch."[16] Thomas alone—officially entered on January 2, 1908—represented the United States at the start of the international Great Race. Other starters and their official entry dates were the French De Dion (November 25, 1907), Motobloc (December 20, 1907) and Sizaire-Naudin (January 18, 1908); German Protos (by January 18, 1908); and Italian Brixia-Zust, more commonly called the Zust (December 20, 1907).

Fig. 2.2 Hans Koeppen of the Protos crew adapted to the cold in his own distinctive style. Here, he is shown in his everyday attire—a German army uniform. (Im Auto um die Welt)

Modified Autos Arrive for Start

"This is more than a race," Hans Hansen of the De Dion proclaimed weeks before the start. "It is an expedition."[17]   Entrants responded accordingly.   Many New York–Paris autos were much stauncher than stock models. At a minimum, automakers reinforced their car's frame; added taller wheels; built a special body for carrying supplies; and installed extra oil, gas and water tanks. Four of the six cars were factory sponsored, and entrants pinned their hopes on cars ranging from 1 to 4 cylinders, 12 to 70 horsepower, and from approximately 3,300 to 6,325 pounds. (See Table 2.1.) The drivers were generally young men—drivers on the American leg of the New York–Paris race averaged 27 years of age[18]—with racing backgrounds. (See Table 2.2.)

Fig. 2.3 For the New York–Paris Race, Koeppen's uniform consisted of furs and a heavy overcoat. (Im Auto um die Welt)

Three contestants—Alphonse Autran (De Dion), Charles Godard (Motobloc) and August Pons (Sizaire-Naudin)—had participated in the grueling 1907 Peking–Paris race that inspired the round-the-world idea. G. Bourcier St. Chaffray, one of the De Dion's drivers, told American newspapers that he helped plan the Peking–Paris contest.

Five cars started in that event on June 10, 1907, four months after *Le Matin* proposed the race of some 9,000 miles. Though one starter dropped out in the early going, the remaining four autos trekked from Peking, through the Gobi Desert and across the boggy tundra and steppes of central Asia to reach Paris, led by Prince Scipione Borghese of Rome in a 40-horsepower Itala. As *Le Matin* prophesied, the race proved that "as long as a man has a car he can do anything and go anywhere."[19]

Traveling on a German ocean liner, the Protos arrived in New York Harbor on Friday evening, February 7, and passed through customs quickly. The remaining four foreign crews arrived Saturday morning. (See Fig. 2.4.) Their cars, however, remained aboard ship awaiting a customs inspection, which began at 7 a.m. Monday, February 10, two days before race time. These four foreign crews gathered at the docks, eager to receive their cars. But a customs inspection meant a day of "long delays and official red tape." The blame rested with race organizers, who could have avoided the inspections by securing "a stroke of the pen from Washington,"[20] according to harried customs officers. After reclaiming their cars that evening, the men hurriedly worked on their machines throughout the night and all day Tuesday.

Table 2.1 Mechanical Details of the New York–Paris Racers

Car:	De Dion (Fr.)	Motobloc (Fr.)	Protos (Ger.)	Sizaire-Naudin (Fr.)	Thomas (U.S.)	Zust (Ital.)
Drive (C = chain) (S = shaft)	S	C	S	S	C	C
Gasoline capacity (gals.)	160–175[1]	72	178	84	125	110
Weight (lbs.)[2]	6,000[3]	6,325	6,000	3,300[4]	4,250[5]	6,000[6]
Cylinders	4	4	4	1	4	4
Horsepower	30–40[7]	24–30[8]	40	12	70	30–40
Weight-to-power ratio (lbs./hp)[9]	171:1	234:1	150:1	275:1	61:1	171:1

[1] According to conflicting reports in the *New York Times*, the De Dion carried either 160 gallons in a seven-compartment tank or 175 gallons in a six-compartment tank.

[2] Given the wide range of reports from the automakers, contestants and press, the weight of the New York–Paris autos is anybody's guess. To further confuse the issue, few reports distinguished between the car's loaded weight (including passengers, supplies, fuel) and unloaded weight. This table lists the mid-range figure given for each car.

[3] This is a compromise: A confused G. Bourcier St. Chaffray, the driver, variously reported the De Dion as weighing 4,500, 6,000 and 7,500 pounds.

[4] From a Feb. 13, 1908, *New York Times* specifications table. A day earlier, the newspaper reported the car's weight as 2,500 pounds.

[5] The *New York Times* originally said the Thomas weighed 3,600 pounds; it later used figures ranging from 4,000 to 4,500 pounds. Montague Roberts told the *Omaha (Neb.) World-Herald* the Thomas weighed 4,300 pounds. In *The Longest Auto Race*, Schuster says the loaded car weighed 5,000 pounds. Perhaps a 4,250-pound figure represents a logical compromise.

[6] Driver Emilio Sirtori once reported the weight as 6,500 pounds; other reports suggest 6,000 pounds is more accurate.

[7] St. Chaffray gave the De Dion's horsepower rating, variously, as 30 or 40, and the *New York Times* did likewise.

[8] Several times, however, the *New York Times* reported the Motobloc as having 40 horsepower.

[9] Dividing the car's weight by its horsepower yields the number of pounds per horsepower—a basis for comparing theoretical performance when autos vary greatly in weight and engine size.

Sources: The *New York Times* and other newspapers.

Table 2.2 Crews of the New York–Paris Autos in the United States, 1908

Key:
D = Driver
M = Mechanic
C = Correspondent
    or reporter
O = Other: guide,
    interpreter, general
    assistance
P = Photographer

Name	Age	Weight	Height
De Dion crew			
Alphonse Autran (D/M)	25	105 lbs.	5-7
Hans Hendrik Hansen (O) (New York City–Chicago)	43	180	5-10
Emanuel Lescares (D) (Chicago–San Francisco)	–	–	–
G. Bourcier St. Chaffray (C/D)	36	130	5-7
Motobloc crew			
Charles Godard (D)	31	176	5-7
Arthur Hue (M)	26	132	5-3
Maurice Livier (D)	19	139	5-6
Protos crew			
Hans Knape (D/M) (New York City–Chicago)	29	165	6-0
Hans Koeppen (C/O)	33	195	6-2
Ernest Maas (D) (New York City–Chicago)	33	176	5-8
O.W. Snyder (D/M) (Chicago–San Francisco)	–	–	–
Thomas crew			
Harold S. Brinker (D) (Ogden, Utah–San Francisco)	22	145	5-10
Mason B. Hatch (O)[1] (Chicago–Boone, Iowa)	–	–	–
Charlie Duprez (P)[2] (Omaha, Neb.–Bakersfield, Calif.)	17	–	–
E. Linn Mathewson (D) (Cheyenne, Wyo.–Ogden, Utah)	23	160	6-1
George Miller (M) (Buffalo, N.Y.–San Francisco)	–	–	–
Montague "Monty" Roberts (D) (New York City–Cheyenne, Wyo.)	24	164	6-1
George Schuster (D/M)	35	–	–
T. Walter Williams (C)[3] (New York City–Clarence, Iowa)	42	–	–
Hans Hendrik Hansen (O) (Omaha–San Francisco)	43	180	5-10
Sizaire-Naudin crew			
Maurice Berthe (M)	24	153	5-5 1/2
Lucien Deschamps[4]	24	163	5-7 1/2
August Pons (D)	32	136	5-3 1/2
Zust crew			
Henri Haaga (D/M)	22	153	5-4
A.L. Ruland (O)[5] (New York City–Ogden, Utah)	–	–	–
Antonio Scarfoglio (C/O)	21	165	5-9
Emilio Sirtori (D)	26	184	5-9

[1] Variously reported as either the sales manager or "race manager" of the E.R. Thomas Motor Company.

[2] Duprez, a 17-year-old passenger, also traveled at times by train and in pilot cars accompanying the Thomas Flyer. Maurice Livier, 19, of the Motobloc, was the youngest official crewman in the race.

[3] He was aboard as the *New York Times'* reporter on the eastern half of the transcontinental trip, but Williams shoveled snow, walked for help and otherwise acted as a regular crewman.

[4] Deschamps waited in vain to get aboard the Sizaire-Naudin at Albany, New York, but never got the chance because the car broke down and withdrew from the race before reaching Albany.

[5] The Zust's American agent, based in New York City.

Sources: "The Starters and Their Cars," *New York Times*, Feb. 13, 1908, p. 4:2; other newspaper accounts; and crewmen's obituaries.

De Dion's "Boxlike Body"

For the round-the-world race cars, form followed function. Entrants were especially concerned about frames and other metal parts becoming brittle in Arctic temperatures. In experiments at the De Dion factory, workers cooled metal with "liquid air" in a vain attempt to find an ideal low-temperature alloy. Instead, the racer would appear with a frame "entirely lined with soft wood to supplement its steel parts, and these are in turn wrapped first with felt and then wound with insulating cloth for the purpose of protecting them from the extreme cold," the *New York Times* said. Many other parts on the special French racer would be "somewhat heavier than those put into the ordinary De Dion car."[21] Few of the cars that arrived in New York City, however, were equipped exactly as had been described in the preceding weeks.

The De Dion's greatest departure from stock was its "towering boxlike body," which made it "one of the most remarkable-looking automobiles ever seen in this city," the *Times* said. (See Fig. 2.5.) "Were it not for the regulation hood, characteristic of the ordinary pleasure car, it would readily be taken for some commercial motor vehicle of special construction.... The body of the car, which is its distinguishing feature, resembles a big crate, its top being considerably above the heads of the men in the front seat. In the centre of the top a wide circular hole has been made in which extra tires will exactly fit."[22] The car carried two men in front, one in back—seated inside the spare tires—and had a collapsible windshield. Two men could sleep under a large cape hood that pulled over the top of the car.

Le Matin and *New York Times'* accounts say the De Dion crew planned to bolt on finned cylinders to transform the engine to air-cooling in Alaska and Siberia, where cooling water was likely to freeze. The car would carry wooden wheels for traction in ice and snow. Georges Cormier and Victor Collignon, who drove De Dions in the Peking–Paris race, helped design the chassis of the factory's New York–Paris car; De Dion founder Marquis Albert de Dion and crewman Hans Hansen designed the body. Reports indicate the transmission and engine—variously reported as developing 30 to 40 horsepower—were stock. After its January 11 completion, the De Dion racer—which used a driveshaft instead of the once-popular drive chains—was "subjected to the severest tests in the Swiss Alps and on the Franco-Italian frontier."[23]

Fig. 2.4 The foreign contestants pose together in New York City. From left to right: 1-Ernest Maas, 2-Hans Koeppen, 3-Maurice Livier, 4-Arthur Hue, 5-Antonio Scarfoglio, 6-Charles Godard, 7-Hans Knape, 8-G. Bourcier St. Chaffray, 9-Hans Hendrik Hansen, 10-Alphonse Autran, 11-Emilio Sirtori, 12-Henri Haaga, 13-Maurice Berthe, 14-August Pons, 15-W.J. Hanley (New York City man who had been slated to drive the withdrawn Maxwell entry), 16-Lucien Deschamps and 17-Fred J. Swentzel (Brooklyn, New York, man who reportedly contracted to drive the Zust from Cheyenne to Paris, but never took the wheel). (Im Auto um die Welt)

"In front of the car is fixed a steel ring and socket, and into it will fit a mast" so the racer could sail over the Siberian tundra, thus conserving fuel. Other equipment would include a hand-operated generator to furnish light at night, "a small but very powerful windlass,"[24] a roller-fed route map and a sextant. A special 175-gallon gas tank was divided into six compartments, any of which could rupture without endangering the entire gas supply.[25]

St. Chaffray Leads De Dion Crew

Joining mechanic Autran and explorer Hansen on De Dion's crew was *Le Matin* representative and De Dion driver G. Bourcier St. Chaffray. Besides helping plan the Peking-Paris race, St. Chaffray "originated a number of automobile races in Europe … and is acknowledged to be one of the most expert drivers in Europe," the *New York Times* reported. He was "slight and short of stature, almost to the point of delicacy, but there is life enough in the quick glance of his piercing black eye." As "Commissionaire General" in charge of the New York–Paris race, he undoubtedly "has acquired a tremendous store of information respecting conditions in Siberia which will be valuable to him in the race,"[26] noted the *New York Times*. St. Chaffray's conflicting role as head of the De Dion crew thus gave him an advantage over other racers. During the race, he telegraphed reports to *Le Matin*.

Fig. 2.5 The De Dion preparing to start from Times Square. Its three crewmen are St. Chaffray, driving, and Autran, the front-seat passenger. Hansen is standing on the "towering boxlike body." The unidentified man in furs is perhaps a guide or other interested party. Visible in the background, far left, is the canvas top of the Protos. (LOC No. LC-B2-23-2)

Autran, a De Dion factory mechanic, served in that capacity on a De Dion that finished second in the Peking–Paris race, according to the *Times*. Hansen, a Norway native and trained engineer, had lived five years in the United States, where he studied railroad bridges in the Rocky Mountains. Hansen had also worked on Mexican and South American railroad projects, according to press accounts.[27] The newly naturalized U.S. citizen told reporters he regretted he could not race an American car; later, he got his wish by joining the Thomas team.

Hansen's resume was full of inconsistencies and contradictions. Many reports said he had spent two years in Siberia looking for a lost North Pole explorer and then 10 years mapping Siberia for the Russian government. The explorer was Major Salomon August Andrée, a Swedish scientist. On July 11, 1897, from the Spitsbergen island group in the Arctic Ocean, Andrée and two companions set off in a hydrogen balloon in hopes of reaching the North Pole by air. They disappeared. The *New York Times* carried a half dozen indexed articles about various rescue attempts but none mentions Hans Hansen's participation. The fate of the three balloonists remained a mystery until 1930, when, "by an almost miraculous chance," another explorer discovered their frozen, well-preserved bodies on White Island near Spitsbergen.[28] Hansen, a "soldier of fortune,"[29] wore an Argentine naval uniform and retained his Captain title from a military stint "in the Argentine Republic, where he served on a warship in one of the revolutions."[30]

Le Matin called upon Hansen to help chart the Siberian route of the race. His knowledge of that route would have given a further advantage to the Frenchmen aboard the De Dion, had Hansen remained with that car. Because Hansen spoke Russian, German, English, Spanish, Chinese and "one or two" Siberian dialects, however, he would prove an invaluable interpreter upon joining the Thomas crew. He was less valuable in other ways, as the tall, blond Hansen "looked like a Viking but didn't know how to drive a car," as Thomas teammate George Schuster wrote in a 1966 book, *The Longest Auto Race*.[31] Confident nonetheless, Hansen bet 10,000 rubles—nearly $8,000—that he could guide the De Dion to Paris within 100 days. "We will either reach Paris or our bodies will be found beside the car,"[32] he declared in the early going. Emanuel Lescares replaced Hansen, who quit the De Dion in Chicago and shortly thereafter joined the Thomas crew.

"Most of these men have not the faintest conception of what they will undergo, and most of them will be left by the wayside before they leave the United States," was Hansen's strikingly accurate prediction. "Our car is equipped with everything—except an arrangement to make it float."[33]

Nervy Godard Drives Motobloc

Motobloc workers started with a "strongly built" and reinforced stock chassis, and then, like the De Dion, lined the frame with soft wood in hopes of combating cold-weather brittleness.[34] The chain-driven French auto—newspapers occasionally spelled it "Moto-Bloc" or "Moto Bloc"—used a 4-cylinder engine developing 24 to 30 horsepower, carried 72 gallons of gas in four tanks and weighed 6,325 pounds. Its special body contained a chest of drawers for spare cylinders and other parts, a cape hood overhead and two double seats. (See Fig. 2.6.)

Charles Godard and Maurice Livier would drive the Motobloc; Arthur Hue was the mechanic. Livier, at 19 the youngest of all New York–Paris racers, was the bioscope operator at the Empire Theatre in London, and a "cine-cameraman," according to a biographer. Livier planned to shoot and send home "enthralling films of the journey."[35] Godard had driven a 4-cylinder, 15-horsepower Holland-made Spyker car in the 1907 Peking–Paris race. Though a lackadaisical planner who compensated with resourcefulness—he borrowed money from anyone who would lend it, for instance—Godard's 1907 performance proved him "the most brilliant automobilist of them all in terms of sheer driving nerve and endurance," biographer Allen Andrews recounts in *The Mad Motorists: The Great Peking-Paris Race of '07*.[36]

When his Spyker sputtered to a halt in eastern Asia's Gobi Desert, Godard persuaded an irregular nomadic cavalry to fetch gas. Though he and *Le Matin* reporter Jean Du Taillis were still very weak from their three-day breakdown in the desert, Godard pursued the leading Itala for days at a time with little sleep. Godard's troubles multiplied nonetheless, as a Paris court sentenced him to 18 months in prison "for obtaining money from the Dutch consular officials

Fig. 2.6 The Motobloc's built-in parts drawers and large gas tank. Translated, the sign reads "World Tour by the Motobloc." (NAHC)

in China by false pretenses." Godard served no time but his arrest near Berlin for extradition to France took him out of the 1907 race. A factory driver took the wheel at Berlin and the car reached Paris with the two De Dions, some two weeks behind the winning Itala.[37] Arriving in New York City for the Great Race, Godard carried "an air of jollity and devil-may-care that is quite in keeping with his reputation for fearless driving,"[38] observed the *New York Times*, whose news accounts occasionally elevated him to the status of "Baron Godard."

German Soldiers Staff Protos

The Berlin newspaper *Zeitung am Mittag* commissioned the German Protos company to build a 4-cylinder, 40-horsepower, shaft-drive auto for the round-the-world race. The crew comprised three German army officers "and it is believed they have orders to go through at no matter what sacrifice."[39] Some 600 factory men helped build the racer in 16 days. It carried 178 gallons in six gas tanks and had a special body wide enough for three men to sleep on the floor. "For this purpose it is covered by a canvas hood that gives it an appearance like that of an army commissary wagon" or prairie schooner, the *New York Times* said.[40] (See Fig. 2.7.) Among the heaviest entrants at 6,000 pounds or more, the Protos blew out its overloaded tires so often that it once forced Lieutenant Hans Koeppen to ride ahead by train to locate a new supply.

Fig. 2.7 The imposing Protos. (Im Auto um die Welt)

The car "can be disassembled with the least possible time and put together again," according to crew member Hans Knape. Knape, who helped design the special Protos, "is a veteran of a half-dozen international automobile contests on the Continent, and drove in the famous Paris-Berlin race in 1901,"[41] the *New York Times* noted. Equipment on the Protos included spades, ice picks, "sledges"—evidently runners or sleds for snow travel—and steel-spiked wheels with which the crew hoped to cross the Bering Strait ice.

Joining Knape and Koeppen on the Protos was Ernest Maas, "one of the leading motorcycle experts in Germany." The three men financed the trip largely by pooling their money. Knape and Maas, both German army engineers, would drive the Protos. Knape, who would double as mechanic, would also be responsible for the "clearing of roads or the making of them, which may be a very large part in the undertaking." Koeppen "is not an automobilist but a sportsman," the *New York Times* said. He took a one-year leave from the German General Staff to accompany the Protos "in the belief that he would have much of military value to report to his army chiefs,"[42] the *Times* said. He would also file dispatches to *Zeitung am Mittag.* Immediately after the race, he also filed a report to the world—his book, *Im Auto um die Welt* (*Around the World in an Auto*).

"Peking" Pons Pilots Pygmy

August Pons, yet another Peking-Paris veteran, would drive the tiny 12-horsepower Sizaire-Naudin. (See Fig. 2.8.) Despite carrying 84 gallons of gas in three tanks, the shaft-driven Sizaire-Naudin was the lightest car in the race. The *New York Times* variously reported its weight as 2,500 and 3,300 pounds loaded—one automotive historian says 4,000 pounds. Though "light" in comparison to the other racers, the 1-cylinder Sizaire-Naudin huffed and puffed up hills in rural New York.

Fig. 2.8 The diminutive Sizaire-Naudin, photographed before leaving France. Pons is at the wheel, Berthe seated beside him. (Im Auto um die Welt)

Pons, father of opera singer Lily Pons, nearly perished in the Gobi Desert when his 6-horse-power Contal tricycle car ran out of gas during the Peking-Paris race. As the cars approached the Gobi Desert early in the contest, Pons' Contal failed to meet the four other autos at the evening camp. Despite an agreement to travel together in the early going, eventual winner Prince Scipione Borghese pressed onward in his fast Itala, declining to search for Pons. The other crews reluctantly followed suit.

Next day, Pons and his mechanic walked 20 miles by 11 a.m., drank from a mud puddle, then decided to walk back to their car, where nomads later discovered them, unconscious in the sand. Recovering, the two men returned to Peking, abandoning their car and the race.[43] Sizaire-Naudin's sole agent for the United States and France, Albert de Laihacar, "a wealthy Frenchman,"[44] sponsored the New York-Paris car. Thus the Sizaire-Naudin, like the Protos, was without factory sponsorship.

Accompanying Pons was co-driver Lucien Deschamps, a "ciné-cameraman," and mechanic Maurice Berthe. Through unexplained circumstances, however, Deschamps "was left behind in New York" and so traveled to Albany, New York, to await the car, the *New York Times* reported.[45]

Thomas Enters Older Model

The day before the start of the Great Race, there arrived in New York City by railcar express a $4,000 bluish-gray Thomas Flyer.[46] (See Fig. 2.9.) The E.R. Thomas Motor Company of Buffalo, New York, chose to enter an older 1907 Model 35 roadster because its 1908 offerings "were not so good on hills," according George Schuster, a driver and mechanic on the New York-Paris Thomas. Even so, Schuster said, the 70-horsepower racer barely got 10 miles per gallon of gas "and it used a troublesome chain drive when many rivals already had shaft drives,"[47] including the De Dion, Protos and Sizaire-Naudin. The Thomas broke four pairs of countershaft housings, a critical part of the chain-drive mechanism, in crossing the United States.

Some latter-day accounts identify the car as a Model 36. But according to *MoTor's 1907 Motor Car Directory*, the Thomas Model 35 was the 3-passenger runabout and the Thomas Model 36 was the 7-passenger side-entrance tonneau. With a bore and stroke of $5\frac{1}{2}$ x $5\frac{1}{2}$ inches, the engines in both models were rated at 60 horsepower. Nearly all contemporary and retrospective accounts refer to the American racer as a 60-horsepower car—but in error.

Fig. 2.9 The Thomas Flyer, America's only entry in the Great Race. (NAHC)

According to Schuster, who in 1964 advised the workers restoring the car, the round-the-world Thomas was a more powerful and relatively rare version of the Model 35, which the factory also designated as its Model D. Schuster said the race car was a Model DX, which differed in two ways from the standard Model D: One, it had a larger bore—5¾ inches, compared to 5½ inches in the other models—enabling the DX engine to produce 70 rather than 60 horsepower. Two, the Model DX had a reinforced frame that was stronger than the Model D's.[48]

America's only entry "is a regular stock car, exactly the same as hundreds of Thomas Flyer owners possess," the Thomas company contended. The car used in the round-the-world race was originally promised to the Whitten-Gilmore Company, Thomas agents in Boston, according to the automaker. It later revealed that it had installed a steel bar between the two front spring hangers—"not … to strengthen the car, but to furnish a suitable attachment for a [tow] rope." The *New York Times* reported, however, that "certain parts have been strengthened for the long journey."[49]

Fitted with two rear bucket seats, the mufflerless car held four people. After modifications on Day 5 of the race, the car carried up to 125 gallons of gas in one large tank, according to most estimates at the time.[50] The car weighed more than two tons—4,300 pounds, according to driver Montague Roberts, but Schuster claimed the car's loaded weight was 5,000 pounds. Some automobile enthusiasts and historians revere the New York–Paris Thomas as the acme of ruggedness—praise appropriate for any car driven 13,000 miles under horrendous conditions. But the men who piloted the Thomas Flyer perhaps deserve the greater credit for overcoming the car's various weaknesses.

Special equipment included an iron-frame canvas top that the crew soon discarded; floorboards perforated to allow engine heat to warm the front-seat occupants; side-mounted skid planks, approximately 16 feet long, for bridging mudholes and small gullies; a 10-gallon oil reservoir; extra springs and other parts; and three searchlights. Upon reaching Cedar Rapids, Iowa, the car carried three rifles and four .45-caliber Colt pistols, all in waterproof cases. "We probably will never need this miniature arsenal," Roberts told the *Cedar Rapids Evening Gazette*, "but there is no telling what we will run into when we get into Alaska and Northern Siberia."[51]

At the midway point of its U.S. transcontinental trip, the Thomas had become a virtual rolling commissary. In contrast to the foreign racers, however, the Americans' assorted supplies added little to the weight of the car. According to *MoTor*, the Thomas carried

> 200 feet Manila rope with pulley blocks; 1 axe and 1 hatchet with extra handles; 2 shovels; 2 snow shovels; 1 crow-bar; 1 pick axe; soldering outfit and 2 pounds of soft solder; 12 files; 24 drills up to ¾ inch; 1 set taps and dies; 1 breast drill; 4 screw drivers; 2 jacks; 4 cans Smooth-On iron cement; 1 package assorted packings; 15 wrenches of various kinds; 50 assorted nuts and bolts; 1 hand lantern with extra wicks; 1 hack saw frame; 4 dozen hack saw blades; 1 wood saw; 2 ball pein hammers; 2 cans valve grinding powder; 100 assorted cotter pins; 100 assorted lock washers; 2 sets of Weed [tire] chains; 2 oil guns; 10 rolls tape; 5 cans of hand soap; 1 folding bucket; 2 sets tire irons; 1 tire patching

outfit; 4 extra tire shoes; 8 extra inner tubes; 12 extra tire valves and miscellaneous; 2 rifles; 3 revolvers; supply of ammunition … leather covering for occupants of car; fur-lined coats; rain coats.…[52]

Three Drivers for Thomas

Three men drove the Thomas across the United States in stages: Montague "Monty" Roberts from New York City to Cheyenne, Wyoming; Thomas dealer E. Linn Mathewson of Denver from Cheyenne to Ogden, Utah; and race-car driver Harold S. Brinker, a Denver native living on the West Coast, from Ogden to San Francisco. Schuster and George Miller, Thomas factory mechanics, accompanied the car, and Schuster also acted as relief driver. Miller joined the crew at Buffalo and often traveled ahead by train to line up guides. Other part-time passengers included *New York Times* reporter T. Walter Williams, who rode from New York City to eastern Iowa; *Times* photographer Charlie Duprez; and Hans Hansen, who defected from the De Dion in Chicago and joined the Thomas at Omaha, Nebraska. Only Schuster, who became chief driver in San Francisco, stayed with the car from New York to Paris.

Roberts, 24, who proposed that a Thomas enter the Great Race, had invaluable racing and engineering experience. For two years while he worked for the U.S. Army Ordnance Department, Roberts, a mechanical engineer, helped build and test "the first gasoline artillery truck ever used by the Government in army manoeuvres," asserted the *New York Times*. This experience "showed me much of value to be used in this contest," said Roberts.[53] As a demonstrator for New York City Thomas agent Harry S. Houpt, Roberts had given driving lessons to Franklin Delano Roosevelt of Hyde Park, New York, a future U.S. president.[54]

On August 9 and 10, 1907, Roberts drove a Thomas 997 miles to win a 24-hour race on the Brighton Beach track at Brooklyn, New York.[55] Many observers gave the colorful, grinning Roberts much of the credit for the Thomas Flyer's strong showing in America. But in speaking of track racing, at least, Roberts downplayed the driver's role: "If I were asked to analyze success in racing, I should say that 50 per cent. was due to preparedness, 30 per cent. to the car, and 20 per cent. to the driver."[56]

Schuster, 35, earning $25 weekly as Thomas' chief road-tester after nearly six years with the company, had his salary doubled for the race. A bicycle racer who worked for a Rambler agent in Buffalo, Schuster in late 1902 joined the Thomas company, where his experience included building radiators, overseeing final assembly and troubleshooting.[57]

Renegade Werner "Racing the Race"

An unofficial entrant, Eugene Lelouvier and his French-made Werner auto arrived in New York City accompanied by "two of the best mechanics in the Werner factory."[58] (See Fig. 2.10.) Lelouvier contended that he conceived the New York-Paris idea but told it to St. Chaffray, who presented it as his own. To retaliate, Lelouvier declared himself an independent contestant who was "racing the race." Thus the self-proclaimed "Sporting Anarchist" left the Pulitzer Building in New York City on February 11—a day before the other cars—to follow a route through Alaska and Canada that the race organizers had found impracticable. That is, from Skagway to Dawson City, and then down the frozen Yukon River through Tanana and eventually striking overland to Nome, on the Bering Sea. Joseph Pulitzer's flagship newspaper, the *New York World*, publicized the rival trip.[59]

Fig. 2.10 With his two mechanics, Lelouvier poses at the wheel of the Werner, fresh from the factory. To his left is Maurice Dreighe and in the back seat is Max Hohmann. (Feb. 1, 1908, L'Automobile)

Zust Carries Italy's Colors

Fitted with a chain drive like the Thomas, the Italian Zust's 4-cylinder engine developed from 30 to 40 horsepower, according to various press accounts. The Zust carried 110 gallons of gas in three tanks and weighed about 6,000 pounds. Customs officials refused to let the Zust crew near its car while it sat for five hours on a New York City pier after it was unloaded from *La Lorraine*, a French ocean liner. The water froze and cracked the radiator, which leaked throughout New York state and beyond. Otherwise gray, the Zust had its hood painted in stripes of red, white and green—colors of the Italian flag. (See Fig. 2.11.)

Emilio Sirtori, 26, "a well-known Italian chauffeur,"[60] drove the Zust and Henri Haaga, 22, was an alternate driver. Haaga, a blond German whose "three months' stay in Milan has made more Italian than the Italians,"[61] was also the mechanic. Another crew member was Antonio

Scarfoglio, 21, son of a Naples, Italy, newspaper publisher. Scarfoglio sent dispatches to London's *Daily Mail*, *La Stampa Sportiva* and his father's newspaper, *Il Mattino*, and afterward wrote a book, *Round the World in a Motor-Car*. A fourth crewman, A.L. Ruland of New York City, an American agent for Zust, rode in the racer most of the way to San Francisco.

The racing teams would follow various strategies. The Zust crew, for instance, "is planning to push ahead with the greatest rapidity," the *New York Times* reported. "The German crew proposes an agreement with the French crew of the De Dion car to proceed more cautiously. The men in charge of the Moto Bloc and Sizaire-Naudin machines believe in keeping together as much as possible as well." Before the start, Roberts said he hoped to average 30 mph to Omaha. He would let road conditions dictate his speed thereafter, he said, adding that the Thomas crew was not making a speed run and had no plans to drive at night.[62]

Confetti and Kisses

Hotels in Times Square were draped with flags and "every window overlooking the square was filled with eager onlookers" on Wednesday, February 12, 1908—Lincoln's birthday and thus a public holiday. Organizers started the contest during the wintertime so the racers could traverse Alaska and Siberia over ice and snow instead of muck and mire. Some 50,000 people crowded around the New York Times building, the newspaper reported. An estimated 150,000 people lined an eight-mile route through New York City and some 50,000 more saw the racers

Fig. 2.11 Haaga (leaning against car) and Sirtori (at the wheel) pose with the Zust in New York City. (AAMA)

pass through the countryside and villages on Wednesday, so that "in all perhaps a quarter of a million people witnessed the first day's progress."[63]

As the six cars approached the starting line (see Fig. 2.12) "a huge American flag, which had been suspended cornucopia fashion from a wire stretched overhead, was opened and down poured a shower of tiny flags, American, French, German, and Italian, a little storm of gay colors."[64] The contestants were at least as colorful. Scarfoglio of the Zust, for instance, "wore a bearskin cap, fur-lined heavy cloth topcoat, and arctic shoes," the *Times* reported. In the Sizaire-Naudin, Pons "was dressed in a blue and black check overcoat and dark tweed cap. His assistants wore flannel caps and rubber coats, while each man was wrapped up in a heavy carriage rug."[65] (See Fig. 2.13.)

Fig. 2.12 From left to right, the cars are the Thomas (American), Zust (Italian), Protos (German), De Dion (French) and Sizaire-Naudin (French). Not pictured is the Motobloc (French), which was next to the Thomas. (NAM)

Because the official starter, Mayor George B. McClellan, was late in arriving, race organizers turned to Colgate Hoyt, president of the Automobile Club of America. McClellan would get his chance to start the New York–Seattle race in 1909. Thus at 11:15 a.m., Hoyt "fired the shot which started the machines on their long race to Paris."[66]

Spectators broke through police lines to crowd against the cars as they crept along Seventh Avenue toward Broadway. (See Figs. 2.14 and 2.15.) A wave of cheers followed the racers, as did a parade of 200 or 300 autos. "I hardly hear the report," wrote Scarfoglio of the Italian Zust car. "I only know that we are hurled forward, surrounded by a crowd, kissed again and again, then advance between two thick hedges of extended hands amidst a roar as of a falling torrent." Recalls Schuster: "Anyone in Times Square that day never forgot the scene, the bands, the flags, the confetti and the screaming, cheering, whistling crowd."[67]

Fig. 2.13 Pons, driving the Sizaire-Naudin, appears to have discarded his blue and black checkered overcoat. But Berthe, next to Pons, has on a rubber coat and is wrapped in a carriage blanket, as described. Visible immediately behind the hood of the Sizaire-Naudin, the De Dion crew is dressed in white. (LOC No. LC-USZ62-55830)

Of Words and Pictures

Newspapers in America, Great Britain, France, Germany, Italy and other European countries wrote thousands of articles chronicling the race. As one of two sponsoring newspapers, the *New York Times* led the way with its race coverage, using telegraphed reports from each car to produce up to a full page of race news every day for six months. Even if stranded on the road at night, contestants were required to telephone their position to the nearest telegraph station. In addition to the three contestants who telegraphed dispatches to other newspapers, the *New York Times* telephoned daily reports to 100 subscribers of its news service.

Newspapers in Chicago, Omaha, Denver (which was off the route) and elsewhere covered the racers' passage for days; smaller newspapers wrote their own stories. In cities along the route, local newspapers and Western Union stations posted race updates for crowds that gathered around bulletin boards. This blanket coverage was an advantage to the racers, who routinely scoured newspapers to learn of their competitors' whereabouts.

Thus the *New York Times* and other newspapers covering the Great Race compiled a vast record—replete with inconsistencies and conflicting information, however. On a single page of racing news, the *New York Times* spelled a foreign contestant's name three ways,[68] for instance, and often rendered George "Schuster" of the Thomas as "Schuester." Elsewhere,

Fig. 2.14 Crowds lining Times Square patiently await the start, held in check by many police officers on foot and horseback. The racers (lower foreground), from left to right, are the Motobloc, Thomas, Zust, Protos, De Dion and Sizaire-Naudin. (Im Auto um die Welt)

Fig. 2.15 Onlookers push forward as the racers disappear into the distance under a pall of exhaust smoke. (NAHC)

the newspaper offered contradictory information about men, machines and mileages. Other newspapers reported differing accounts of daily events, and three crewmen who wrote books afterward—Koeppen, Scarfoglio and Schuster—introduced inconsistencies that further dilute the record. Unfamiliar with English, U.S. geography and automobile mechanics, some of the crewmen-reporters aboard the foreign cars left their readers wondering what broke and where. All of this makes the truth of what happened during the 1908 New York-Paris race surprisingly elusive.

There were attempts made to capture the Great Race on film, as well as in words, but it's unknown whether any such footage survives. The American Vitagraph Company, which filmed the start of the race, began showing its footage several days later at Hammerstein's Victoria Theatre of Varieties in New York. "The film is pronounced entirely satisfactory, and is said to show all the important features of the ceremony most distinctly,"[69] according to the *New York Times*.

Maurice Livier had planned to film his view of the race from the Motobloc—and left Chicago with a movie camera "perched on a specially contrived rest, over the hood."[70] On Day 27, March 9, a motion-picture photographer left Cheyenne ahead of the Thomas and filmed the auto coming through a canyon on its way to Laramie, Schuster recalled. But "I never saw or heard anything of this afterward, or of any moving pictures taken of the trip,"[71] he said in later years.

Either this 1908 contest or the New York-Seattle race that followed in 1909 may have inspired *The Race*, a 1916 silent film about a transcontinental auto race. It starred Anita King, who actually drove an auto across the country in 1915. In *The Great Race*, a 1965 movie, Tony Curtis, Jack Lemmon, Natalie Wood and others spoofed the intense rivalries that developed among crews during the 1908 contest.

Three Cars Take Lead

Over snowy roads and two-foot drifts, the De Dion, Thomas and Zust pulled ahead of the other autos and generally traveled together during the first five days. They reached Hudson, 116 miles from New York City, the first night, and spent succeeding nights at Fonda (90 miles), Canastota (71 miles), Geneva (77 miles) and Buffalo (119 miles). The Motobloc, Protos and Sizaire-Naudin trailed. After a falling-out at Buffalo, the three leading crews parted company and the Thomas opened up a tenuous lead into Chicago. As for the first night, however, the Thomas arrived in Hudson at 8:20 p.m. The Zust arrived more than an hour later, 15 minutes ahead of the De Dion. This was the order, following their breakup, that the three autos would maintain in the race across America.

"One difficulty the drivers experienced," *New York Times* reporter T. Walter Williams in the Thomas wrote on Day 1, "was the fashion country people have of walking in the middle of the road and refusing to step aside."[72] But the six New York-Paris racers experienced other delays in the early going. At Peekskill, New York, on Day 1, the Zust's engine stopped suddenly, overheated. The radiator was still leaking and had lost all its water, which the crew replaced by melting ice in the radiator, Scarfoglio wrote.

Brake troubles forced the Protos to halt at Poughkeepsie after 74 miles. All day the 1-cylinder Sizaire-Naudin "found the hills very difficult to negotiate with the low power it had."[73] Late in the afternoon after traveling 40 miles, the Sizaire ground to a halt, apparently with differential problems, on a hill near the suitably French-sounding Montrose, outside Peekskill. The exact location was later reported as Crugar's Station. The French crew slept at a nearby house.

At 9:30 p.m., the Motobloc passed Peekskill, 44 miles from New York City, but stalled in a four-foot snowdrift five miles north of town. Godard and his crew dug out and, after midnight, returned to Peekskill, where they found rooms for the night but no food. It was their day for delays. Earlier at Ossining, as mechanic Arthur Hue cranked the Motobloc, the starting handle kicked back and nearly broke his wrist. A doctor diagnosed the first injury of the New York-Paris race as a severe sprain and bandaged it. Further delays came when a Dobbs Ferry innkeeper kept the Motobloc crew waiting three hours for lunch, hoping to profit by having a team of famous New York-Paris racers in his restaurant, Godard charged.

Rebels Wreck

Meanwhile, the Werner car that was "racing the race" by heading south out of New York City on a longer round-the-world route, was broken down in Philadelphia. A day before starting, the Werner's crew used rope and wire to wrap the car's springs, "so that if one breaks it will hold together until repairs can be made,"[74] the *New York World* said. According to mechanic Maurice Dreighe, the 15-horsepower Werner, whose weight was given as 3,000 pounds but was probably heavier, carried 75 gallons—enough gas for 1,000 miles. The engine evidently had 4 cylinders, though this is never made clear. The Werner crew chose "heavy French tires with brass riveted treads" and stowed seven spares on the side of the car.

Besides tools and extra parts, the Werner carried a coffee pot, frying pan, drinking cups, tin breadbox, water bottles, ax, snow shovel, thermometer, barometer, compass and sextant. "The men will sleep in a huge fur robe when they are unable to find shelter. The members of the other New York to Paris automobile party … think that goggles, shotguns and the like are necessary, but M. Dr[ei]ghe says he doesn't need them,"[75] the *World* said.

Dreighe, who "has spent four years exploring Alaska," spoke English. So did Max Hohmann, the other Werner mechanic, who "was a trapper in Northern Canada in 1893, a cowboy in Montana and Oregon, and is now quartermaster in the submarine branch of the French navy. He, too, has been around the world." The other globetrotter on the Werner crew was Eugene Lelouvier, who spoke little or no English. He was hired as a mechanic for Georges Cormier's De Dion in the Peking-Paris race, and spent five weeks in an ox cart before the race, charting the best route north to Irkutsk, Siberia. For unspecified reasons—possibly a personality clash—

Cormier fired Lelouvier before the race began, Allen Andrews wrote in *The Mad Motorists*.[76] Shipwrecked twice as a merchant mariner, Lelouvier could—and did—boast of walking around the world through Siberia. There he met his future wife, who helped arrange his Siberian fuel supply for the 1908 race.[77]

The Werner crew left the Pulitzer Building a day before the official New York–Paris autos departed, to follow a southern U.S. route through Philadelphia, Pittsburgh, Cincinnati, St. Louis, Kansas City, Cheyenne, Pocatello, Portland and Tacoma to Seattle. Their long day of mishaps—which began not far from the starting line—presaged a difficult circumnavigation. At 1:31 p.m. Tuesday, February 11, some 20,000 New Yorkers witnessed the Werner's start, estimated the *World*, whose reporter joined the crew "for a short distance."[78] But it took the men 16 hours to cover the 96 miles to Philadelphia.

Werner Crew Ejects Lelouvier

"They were snowbound in deep drifts twice and held up for hours by the pranks of the automobile's machinery," the *World* said. "Before reaching the Staten Island ferry the Werner car balked, but it was finally persuaded to crawl aboard the waiting municipal ferryboat." At Stapleton, New York, on Staten Island, "the car bumped into a pony cart. At St. George a few nuts flew off and it was two hours and eight minutes before the car moved again. In Perth Amboy M. Le Louvier, who kept the wheel, complained that the brake was working badly. At any rate a grocery wagon in the street was shoved to the sidewalk.

"Five stops for repairs were made between New Brunswick and Trenton [New Jersey]. At the latter place kindling wood was made of a picket fence separating the road from a ditch. Ten miles from the village of Tulleytown the car skidded while going twenty miles an hour and landed stern foremost in a snowdrift." The car was still stuck four hours later, at 2:15 a.m. Wednesday, when two Philadelphia autos arrived. "They stopped and for three hours towed the Werner car in turn," the *World* recounted. "Once when they let loose it dove into a snowbank again."[79]

The Werner car lingered in Philadelphia for 2½ days with a broken clutch. As the delays mounted, the *World's* coverage dwindled to one-paragraph dispatches from Philadelphia, suggesting its reporter had not ventured to travel so far. After four days of mishaps, the Werner crew mutinied and ejected Lelouvier, the former sailor, from behind the wheel. "Lelouvier was the driver of the car, but last night, following several accidents, his teammates told him he would have to give up control of the steering wheel or stay behind," said a February 15 wire report. "Lelouvier said that if he couldn't drive he wouldn't go at all," and got off at Norristown, Pennsylvania.[80] The *World* soon fell silent on the whole matter. A rare paragraph or two in the *New York Times*, however, tracked the Werner's misadventures across America to Seattle.

Yet another car would share the limelight with the international racers. Sponsored by the *New York American* and attracting shirttail publicity, a Studebaker "U.S. Army Dispatch Car" left

New York six days behind the official racers, attempting to show the automobile's utility in delivering military messages. (See Fig. 2.16.) Operated around the clock by relay drivers, the car reached Chicago the same day as the Thomas. The Studebaker turned off the New York–Paris route at Omaha, heading south to end an 18-day, 1,600-mile trek on March 7 by delivering its message to Fort Leavenworth, Kansas.

Fig. 2.16 The Studebaker Dispatch Car leaving the New York American offices with an unidentified crew. Though obscured by fresh paint, the number "14" on the gas tank suggests that the Studebaker was formerly a race car. (AAMA)

Dashing through the Snow

For the six official New York–Paris racers, snow and New York's steep hills—which held the racers' speed to a few miles per hour—were by far the greatest obstacles. The race degenerated into a shoveling match for the first 1,000 miles. Planning to travel together as far as Chicago, crewman aboard the leading De Dion, Thomas and Zust cars pitched in to help push or shovel whenever one of the cars bogged down.

"Oh, yes, I win the race!" quipped Zust driver Emilio Sirtori. "I shovel much snow." Inside, however, he was raging. "My heart is full of evil thoughts about the men who make the roads," he confessed. "No roads in Alaska or Siberia can be worse than these here in New York State." After just two days on rural New York roads, Hans Koeppen of the Protos agreed. "I expect Siberia will be a picnic compared with what we have experienced to-day." The low-slung De Dion had more trouble than the others in bucking drifts. St. Chaffray, exaggerating only slightly, said: "We are workmen, we are miners."[81] *Chicago Daily Tribune* cartoonist C.A. Briggs showed a neck-and-neck race between Italian, German and French cars—each plowing through the snow behind a horse. "Sacre bleu," the French driver mutters. "Why deet I leef zee gay Par-ee."[82] (See Fig. 2.17.)

Expecting to find American roads equal to those of Europe, most of the foreign drivers had been looking ahead with dread to Alaska and Siberia. They were consequently unprepared for and, with the exception of the Sizaire-Naudin, too heavy to traverse foot-deep mud, huge

Fig. 2.17 Cartoonist C.A. Briggs' version of "The Road to Paris."
(Feb. 27, 1908, Chicago Daily Tribune)

snowdrifts and the unmarked, meandering trails that Americans called highways. "All five of them were equipped with enough spare material to build another car or start an iron foundry," Williams observed in the *New York Times*, overlooking the small Sizaire. "All this made a difference between them and the lightly equipped Thomas car."[83]

Cars Lighten Loads

Such bad conditions, however, surprised even the Thomas crewmen, who were expecting tough going from the start. "The roads up State are simply atrocious, and all the drivers are disgusted," Roberts said in rural New York state. "Even the main highways are blocked up with snow in places 10 to 12 feet deep, and no attempts are made to clear a track."[84] The Thomas carried relatively few tools and parts "due to its crew's determination to rely on local aid in the case of breakdowns," according to automotive historian T.R. Nicholson.[85] Even so, reporter Williams revealed on Day 9 that Schuster on the Thomas "is so absorbed in the race and keeping the lead that he refuses to accept any silver money in change, so as not to add to the weight of the car."[86]

As equipped at the start, the De Dion, Motobloc, Protos and Zust each weighed at least 6,000 pounds. The exact weight is impossible to know because the contestants and newspapers offered so many conflicting figures for each vehicle. The Thomas, however, weighed some 1,500 pounds less than the heavy foreign cars. Lightest of all at approximately 3,300 pounds, the Sizaire-Naudin was nevertheless too heavy for its 1-cylinder engine.

The advantage of a light car soon became apparent. The heavy foreign cars wheezed up hills, regularly popped their tires and got stuck in the snow where a lighter car could pass. On "several occasions" east of Syracuse, New York, the German Army officers had been forced to unload the Protos, drive it over a bad spot, then walk back to retrieve their baggage.[87] They popped four tires in as many days. Before leaving Utica, New York, on the fourth morning, they shipped 600 pounds of supplies ahead to Chicago.

Similarly, the Zust will be "relieved of some of its heavy packages when it reaches this city," Williams reported from Buffalo. "There is a vast difference between a car weighing 3,700 pounds and one tipping the scales at 6,500 pounds," fumed Sirtori, the Zust driver, quoting an optimistically low figure for the weight of the Thomas.[88] In Buffalo, Hansen reportedly sawed off the top of the De Dion's "towering boxlike body" and installed a canvas cover, saving 800 pounds. The Motobloc, too, lightened its load, and to climb the Rocky Mountains even the Thomas shipped luggage ahead. But some insistent spectators made it difficult for the leaders to travel light: On Day 6 between Buffalo and Erie, Pennsylvania,

> [a]s the cars came fast over the snow-covered frozen ground, the crowds cheered lustily and threw all kinds of things into the laps of the contestants, including bananas, apples, oranges, nuts of all kinds, and one generous person hurled a bottle of port wine into the Thomas car and shouted "Drink to that, boys, for luck."[89]

Primitive Antagonism Toward Strangers

The racers had to surmount other difficulties besides bad roads and overweight cars. "The monotony of the long ride in the cold air was broken at frequent intervals by small boys throwing snowballs at the chauffeurs," Williams reported on Day 1.[90] Worse, a policeman confronted the three leading teams as they walked to their cars in Hudson at 6:30 a.m. on Day 2, demanding $3. The previous night a speeding auto caused a horse to jump into a ditch and damage a sleigh, the officer said. All three drivers "were positive that they had not seen a human being when they reached Hudson on the preceding evening." They pleaded their case for an hour; the policeman insisted on their guilt. To escape him, the De Dion, Thomas and Zust crews each contributed $1.

The same day at a small town south of Albany, the Thomas crew met an honest-looking "kind old man" who professed "I love to see you tearing over the country in those great machines of yours."[91] The man purposely misdirected the Americans, who consequently spent an hour digging out of a snowdrift, Williams reported in the *New York Times*. Hoping to make better progress, the leading cars on Day 2 began following the frozen Erie Canal towpath, used by horses and mules to pull boats by ropes. Two days later, a man whose ice-cutting operations blocked the towpath demanded $2 apiece to let the New York–Paris leaders pass. The De Dion paid; the Thomas and Zust crews refused. Only after a "very heated argument" did the iceman move his equipment to let them pass, Roberts recalled.[92]

The racers were encountering "people in whom still lingers the primitive antagonism to the stranger," the *New York Times* said in a chiding editorial. But the *Times* was more distressed by the state's primitive highways. Until the racers leave New York, "its lack of one of the best proofs of civilization—good roads—is going to get daily, wide, and effective advertising."[93]

Towpath or Turnpike?

On Day 3, the leaders had to roll 80 big logs off the icy, muddy Erie Canal towpath and build a road across a deep ditch. As the cars neared Utica, the Zust suddenly zoomed ahead. Sirtori evidently wanted to impress the 10,000 waiting spectators, of whom some 3,000 were Italians, the *New York Times* estimated. He pushed the Zust to the "terrific pace" of 45 mph on the high and narrow towpath between the canal and Mohawk River. The De Dion let the Thomas car pass to pursue the Zust. "Life is too short to go at that pace in such a dangerous spot," explained De Dion passenger Hans Hansen, "and our car will take things more calmly." But it was wishful thinking, for De Dion driver St. Chaffray reacted to the Thomas-Zust speed duel by declaring: "I am making a race now."[94] Though Roberts sped after the Zust, he later complained that Sirtori was driving 15 to 20 mph above a safe speed.

But generally, the cars were reluctant to lead over the snow-clogged roads. "Roberts, of the Thomas, said that his car had been banging into snowdrifts to make tracks ever since they had left New York, and it was a heavy strain on the engine," the *New York Times* reported on

Day 3, when the De Dion and Zust crews agreed to share the trail-breaking task. Ramming through the snowdrifts also strained the Thomas car's radius rods, designed to keep the drive chains tight. "After each bad drift, I had to get underneath with a jack and straighten them," mechanic George Schuster recalled.[95] He phoned ahead to have the Buffalo factory make a stauncher pair.

The racers got little sleep during their full days. As the *New York Times* reported, "The ordinary routine is call at 5 a.m., breakfast at 5:30, go to the garage to make ready, put the baggage on, and leave the hotel at 7, stop for half an hour midday, travel to 7 or 8 in the evening, wash, dine, then go to the garage for a couple of hours, and turn in at midnight."[96]

Cabbages and Cooperation

At times during the daily struggle over snowbound roads, Schuster would leave the Thomas to walk into adjoining fields, using a stick to test the snow depth, wrote Scarfoglio of the Zust:

> When it appears to be too deep he comes back and explores, looking for a better place to cross the fields. He breaks a wire fence, and the cars advance across cultivated fields, bumping across invisible ditches, oscillating, staggering, stopping, then going on again, always following the zigzag tracks of Schuster....
>
> Terrified countrymen watch us pass through the midst of their cabbages, breaking through the boundaries of their fields. They try to protest, but we pay no attention. We quickly become accustomed to this absolute contempt for property, this violent exercise of our will. From this to robbery is only a small step![97]

"There have not been any disputes among the crews of the cars as to who should shovel snow, chop down fences, or build bridges. All hands have turned to willingly," reporter Williams wrote on Day 4. The helpers included Williams and Zust agent A.L. Ruland. "Idlers would not be popular," added Williams, a former seaman who thus became a reluctant racer.[98] A native of England who reported on the Boer War of 1899–1902 for other newspapers, Williams joined the *New York Times* in 1905. As ship-news reporter and marine editor of the *Times* during the ensuing 37 years, he earned the nickname "Skipper" Williams. His command of French, German and Italian made him an ideal choice to cover the New York–Paris race.[99]

On Day 4, from a place Williams called Dismal Hollow on a route through north-central New York's Montezuma Swamp—"famous as being the worst road known in the United States"[100]—Ruland of the Zust auto walked $1\frac{1}{2}$ miles to hire six horses. They pulled the De Dion and Zust from the mire; the Thomas, which avoided the mudhole that claimed the other two cars, got through unaided, according to Williams.[101]

The oldest New York-Paris crewman at 43, Hansen was also the strongest shoveler, according to St. Chaffray: "'After you, Captain,' is the cry of the men of the other cars, as they courteously step aside and let him exercise his muscles on the hardest work." Williams was astonished "that no petty jealousies have developed among the crews of the cars." Still, the wary racers hesitated outside a Geneva garage late in the afternoon on Day 4. "They were afraid to be the first to put their machines in the garage, in case the others went ahead," said Williams, who broke the impasse by questioning each driver in turn. Each one said of the others: "I'll stay if they will," so they stayed.[102]

Mishaps to Zust, De Dion

This tense cooperation ended on Day 5, February 16, when the idea that the transcontinental crossing would be an orderly endurance contest "was speedily dispelled by the Zust car taking the lead at a great pace, so that the Thomas had to put on speed to overtake her," Williams wrote. "From that time each car went ahead as fast as possible when there was a chance."[103] But Sirtori paid dearly for his speeding. On Day 5, the Italian car skidded at the foot of a hill southeast of Rochester, and landed in a snowbank. The crash broke a radius rod—the same part that proved too weak on the Thomas car. Between New York and San Francisco, Italian-Americans couldn't do enough to help the Zust. "They pay everything where we pass, and don't allow us to buy even a box of matches," Scarfoglio said.[104] In this instance, however, it was the Americans at the Gearless Transmission Company factory in nearby Rochester who came to the rescue.

Gearless workers helped as the Italian crew attempted to forge a new radius-rod knuckle. But the factory's tooling allowed them to only "perfect a rough substitute," said Sirtori, who had the part reinforced in Chicago.[105] The automaker sent one of its Gearless cars to act as the Italians' guide and trail-breaker all the way to the Windy City and a ways beyond, as it had done in late 1907 to direct record-setting transcontinental walker Edward Payson Weston.[106] "I consider it one of the finest examples of sportsmanship, hospitality, and good fellowship I have ever experienced, and America should be proud of them," a grateful Sirtori responded.[107]

The De Dion also had an accident on Day 5. At one point between Geneva and Buffalo, St. Chaffray tried to turn the De Dion's steering wheel. It wouldn't budge. The car plunged off the road into a field, pitching Hansen into the snow. They discovered "a mass of frozen mud and ice" clogging the steering gear.[108] Back on the road after three hours of shoveling, the men freed the steering mechanism and continued.

Goodwill Going, Going, Gone

Next day, the Zust sped through Buffalo, home of the Thomas factory, while the De Dion and Thomas crews slept. The Zust had finished 12 hours of repairs at the Gearless factory in Rochester, and then started for Buffalo after midnight on February 17—early in Day 6. The three leaders had agreed to attend a Buffalo Automobile Club banquet that night. The Zust somehow missed two Thomas cars that had driven east to escort the Italians into Buffalo, where the streets were nearly deserted when the car pulled in at 5:30 a.m.

As the Zust stopped to telegraph a report to the *New York Times*, "the usual early morning cheerful idiot came up to the car" and erroneously reported that the De Dion and Thomas had set out for Erie, according to Williams. So the Zust crew ate breakfast and left. Telegrams were sent ahead to have the Zust return—a car also pursued the Italians, according to Roberts—but "suddenly St. Chaffray decided to take the De Dion car out of the garage, and started for Erie," prompting the Thomas crew to follow.[109]

The *New York Times*' coverage, Sirtori's telegraphed report from Buffalo and Scarfoglio's book all suggest the Zust crew made an honest mistake. The Thomas crew argued otherwise. "They refused to come back when the car overtook them, and we learned this at 11 o'clock in the morning," Roberts asserted. "Then without warning [the] De Dion started out in pursuit, and I, outside of the town at the [Thomas] factory, did not learn they were gone until 2 o'clock in the afternoon." The Thomas crewmen, suddenly in third place, packed quickly and left town about 3 p.m. In his 1966 book, Schuster contends that the Zust crew sped out of Buffalo after leaving behind the message: "We will await you in San Francisco."[110]

Despite its head start, the Zust lost its lead when it stopped at Ripley, New York, for repairs. That night, at a shared dinner in Erie, Roberts reacted to "several taunting remarks" from St. Chaffray, Schuster wrote in *The Longest Auto Race*. Roberts finally snapped at the French driver, "From now on you will know this is a race!" Hansen, with the De Dion, recalled it differently for the *Chicago Daily Tribune*, saying that St. Chaffray, as commissionaire general of the race, issued Roberts an order: "When you wish to go into a city ahead, you ask me." Roberts angrily replied: "Do you suppose that I am going to run a hippodrome race? I'll lead you all."[111]

Of the incident in Erie, Roberts revealed only that "what I told him would not look well in print, and in the future the more miles there are between us the better it will suit me."[112] Regardless of how it actually happened, international goodwill—in the form of friendly cooperation among racers—had disintegrated. Thereafter, Roberts relentlessly sought to widen his lead, and the casual wintertime jaunt to Chicago became a desperate day-and-night race.

Trouble Behind

Where were the trailing cars at the end of Day 6? The Protos, in fourth place, had reached Buffalo, 473 miles from New York City by the route selected. To get there, the Germans had to beg gasoline from two men in a Cadillac runabout. At Buffalo, the Protos went directly to the automobile factory of the George N. Pierce Company, and remained there until the following afternoon. "There are a few repairs to be looked after, occasioned by a bad rut in the road a few miles west of Batavia," the *Times* said. Its chronic tire troubles had also slowed the Protos. The Motobloc, plagued by a number of small breakdowns, reached Geneva in fifth place on Day 6, 352 miles from Times Square. "I travel far, not fast," a laconic Godard observed.[113]

A crewman aboard the 1-cylinder Sizaire-Naudin had taken the train to New York in search of a new differential. With parts in hand, he returned late on Day 2 to the disabled car at Crugar's Station, near Peekskill. The following day, the crew discovered a piece of iron had wedged in the Sizaire's differential gear, which didn't need replacing after all. But Pons and crew "determined to overhaul the car completely, and the rear wheels were taken off, the axle changed, and new wheels and tires affixed" on Day 4, before the men drove four miles to Peekskill for the night.[114] The car gained 52 miles on Day 5 despite worsening engine troubles that prompted a stop at Red Hook, where the car remained on Day 6.

"The trouble developed in the cylinder, but its exact character has not been yet determined,"the *New York Times* reported. On Day 7, February 18, driver Pons returned by train to New York City "and announced that it [the car] was hopelessly disabled. A part of it was evidently cast with a flaw in the steel, and this part had broken and cannot be replaced."[115] News reports failed to name the flawed part. The smallest car in the race thus bowed out after 96 miles. Lucien Deschamps, who was waiting in Albany to board the Sizaire-Naudin, never got a chance to take part in the Great Race.

Thomas Takes Toledo

On February 18, the day after Roberts and St. Chaffray quarreled, the Americans left Erie at 7 a.m. Propelled by Roberts' anger, the Thomas traveled 219 miles to reach Toledo, Ohio, at 9:35 p.m.—the best daily performance of any New York–Paris racer to that point. The race moved some spectators in mysterious ways. "One commanding-looking woman stood on the porch of her house, near Unionville [Ohio], with her four children, and fired a salute of sixteen shots with a rifle as the American car went by," Williams wrote. "Roberts smiled and waved his hand to her."[116]

Later, he observed, the Americans raced through Willoughby, Ohio, at 35 mph. "The police seemed to have relaxed the traffic laws for the occasion." It snowed all evening, making Roberts' task all the harder. "My arms are nearly pulled off with the strain of driving to-day such a distance over rough roads," he wrote from Toledo.[117]

Starting later and stopping sooner, the second-place De Dion traveled 190 miles to Fremont, Ohio, arriving shortly before the dining rooms closed. "We are heroes, but we eat," proclaimed St. Chaffray, who repeatedly denounced the Americans for turning an endurance contest into a race: "The Americans do not like their own hotels, I suppose, for the men in the Thomas car never stop for luncheon. They seem to look at these houses as bad because they are not moving as fast as a flying car."[118]

St. Chaffray undoubtedly felt justified in his anger. After all, the racers had earlier agreed that the eastern U.S. leg would be "merely a tour," as Roberts conceded. But that was before the disputes and personality clashes. "It has now narrowed down to a race," Roberts declared, a truth St. Chaffray was slow to grasp.[119] Under the rules it *was* a race—the first car to reach Paris, or come closer than the others, would win.

Like the De Dion and Thomas, the Zust entered Ohio on February 18, reaching Cleveland for the night in third place. At an unnamed factory in Erie, the Zust crewmen removed the damaged radius rod to attempt another repair. "Just as we were making good progress the engine driving the machinery broke down," and the work had to be finished manually by lamp light, Sirtori wrote.[120] Between driving and repairing their car, the Zust crewmen had been awake 60 hours.

Racers Go Despite Snow

The weather in western Ohio and Indiana became even less cooperative than the contestants aboard the three leading racers. In Indiana, 10- and 12-foot drifts clogged the highways, and some farmers had gone a week without mail deliveries. But under pressure to lead, the New York–Paris racers bullied their way through by hiring dozens of snow-shovelers and horses. On February 19, Day 8, the Zust and its Gearless pilot car left Cleveland at 1 a.m. and bucked snow for 18 hours until the other crewmen persuaded Sirtori to stop for rest at 7 p.m. in Bryan, Ohio. The Italians would resume their trek shortly after midnight, however.

Eighteen inches of snow blanketed the ground in Toledo on February 19. According to Williams, the snow prompted the leading Thomas car to wait for the De Dion—an improbable story given the enmity between the Thomas and De Dion crews. In *The Longest Auto Race*, Schuster contends that Williams actually overslept, which allowed the De Dion to catch up. Led by a pilot car, the two racers traveled together into northeastern Indiana. The storm worsened near Waterloo at sundown. Shortly thereafter, the De Dion hit a snow bank and bent what St. Chaffray called its "speed shaft"—actually, a transmission shaft. Horses towed the car into Corunna. Next day, the De Dion limped into a larger city several miles west, Kendallville, where Hansen and Autran, with the damaged part, boarded a train for Chicago. There, they had the part mended, and returned to Kendallville to resume racing on Day 12 after three idle days. The Thomas went on after the De Dion's accident. When the Americans got stuck in a deep drift near Kendallville, "all the farmers in the neighborhood helped to dig the car out," Williams wrote.[121]

Beginning the next day, February 20, Day 9, the howling blizzard forced the leaders in the New York–Paris race to frequently abandon the drifted-in roads and drive over farm fields and railroad tracks. The Thomas crew pushed and pulled its car nine miles on Day 9; the De Dion racers crawled seven. (See Fig. 2.18.) Failing to anticipate an Indiana blizzard, the rules of the New York–Paris race were silent on the question of following railroad tracks. Common sense suggested that the practice would at least violate the spirit of the contest. But every crew bent the rules in Indiana by driving on railroad tracks to avoid clogged roads;

hiring snow-shovelers; and using horses to break a trail, pull its car from drifts, or tow the auto for longer distances. (See Fig. 2.19.) Later, the various teams quarreled about who benefited most by such tactics.

The Thomas arrived in Kendallville at 8:30 a.m. February 20 after an all-night effort. "There was no glory or hurrah in the trip last night," Roberts reported. "It was simply a continuous snow-shoveling and pulling and hauling performance." The racers and their helpers, including hired farmers, had to dig almost constantly, Williams wrote in the *New York Times*. "The automobilist who walked ahead to explore the road frequently fell into drifts up to his neck, and had to be dragged out by his comrades. The automobiles were taken across fields to avoid the snow, but as soon as one was cleared another snowbank was sighted right ahead, most of them from 130 to 500 feet long."[122]

After a short rest in Kendallville, the leading Thomas car set off again but averaged barely more than 1 mph during a two-day, 35-mile drive to Goshen, Indiana, on February 20 and 21. Compounding the problem, there were no road signs and "the Hoosiers did not seem to have any knowledge of anything more than half a mile from their own homes," Williams reported. Thus the Americans followed signs that began 50 miles east of "Buggins's Shoe Emporium," situated in an Indiana town Williams fails to name:

> The signs were repeated every two and a half miles, until the contestants felt a positive interest in the personality of Buggins, and wanted to meet him. When the store was

Fig. 2.18 Helpers break out shovels to aid the half-buried Thomas in Indiana. Roberts is at the wheel, Schuster beside the car. (LOC No. LC-USZ62-68652)

Fig. 2.19 Two horse-drawn sleds break a trail for the flag-flying Thomas in Indiana. (NAHC)

finally reached Roberts called out through his megaphone to a prosperous man standing at the door, "Say, Mister, are you Buggins?" The man nodded in the affirmative, and Roberts waved his hand and said: "That's good enough, old man. We just wanted to see you. That's all."[123]

Train Tracks and Horses

Enough was enough, Roberts fumed. The next day, February 22, with the railroad's permission, the Americans followed a special car along the Chicago, South Bend and Northern Indiana Railway tracks. (See Fig. 2.20.) Bumping over the ties, they traveled 28 miles from Goshen in three hours, then finished the remaining 14 miles to South Bend on the road through "the deepest snow yet encountered."[124]

The Zust followed on the tracks a day later—without permission. The officials authorized to give such permission were out of town, the railroad superintendent told the *New York Times*. "The road, however, did all it could to help along the Zust when it finally took to the tracks," the newspaper added. "The auto was met by an interurban car and a report was sent back to the dispatcher who issued orders to let them remain on the tracks, and to inform the train crews to be careful not to run them down."[125]

The railway superintendent on February 23 likewise denied permission to the De Dion, telling the French crew that only the American car would receive the privilege, De Dion representatives alleged. That night, the De Dion slipped onto the tracks, narrowly escaping an oncoming express train before reaching South Bend at 5 a.m. Earlier, the Lake Shore and Michigan Southern Railroad showed more courtesy by ordering its Lake Shore Limited to make a special stop at Kendallville. This saved Autran and Hansen many hours in returning

Fig. 2.20 The Thomas follows a rough but passable roadway between Goshen and South Bend, Indiana. Motor Age *identifies the stream as Salt Creek. (SMI)*

from Chicago with the De Dion's repaired transmission shaft. Godard drove the Motobloc along the same section, presumably without permission since he made the run between midnight and 6 a.m. on February 27. The Protos also traveled on the trolley tracks for part of its journey between Goshen and South Bend.

After leaving the railroad tracks and reaching South Bend by road, the Thomas headed west on the hilly road to New Carlisle and Michigan City, Indiana. It took six horses to haul the Thomas up one hill. Soon, two more teams hitched on. Finally, 12 horses were pulling the car up the steep hills. "Roberts was very averse to using horses to pull his automobile, but he realized that he could not get through to Michigan City in a month while the snow lasted," wrote Williams.[126]

Trailing in the Zust and unaware of the deteriorating roads ahead, Sirtori retorted that he had no intention of hiring farm horses to pull his car, for which its Gearless pilot car was clearing a trail: "The big [Gearless] machine would dash into a mountain of solid snow, throwing it in every direction, and then back up for another assault. In this way the road was broken and the advance of the big foreign machine made possible," the *New York Times* reported.[127]

But a day after printing Sirtori's retort, the newspaper reported without fanfare that the De Dion, Thomas, Zust and the Zust's pilot car "are being hauled through the deep snow by farm

Fig. 2.21 "The Great New York to Paris Automobile Race."
(March 4, 1908, Omaha Daily News*)*

horses."[128] Cartoonists had a field day. (See Fig. 2.21.) Defending the use of horses, President Erwin R. Thomas of the Thomas company cited from the rules that he said the *New York Times* had given contestants before the start: "That as far as possible all the cars shall proceed by the route outlined under their own power, but that assistance may be secured when essential to extricate a car from difficulties."[129] The Thomas car again took to the tracks between Burdick and Hobart, Indiana, for a distance similar to its first run along a railroad right-of-way. It was able to drive from Hobart to Chicago unaided.

The foreign crews, the Chicago Automobile Club, the *New York Times* and many Americans were incensed that foreign crews had been denied permission to travel over the Indiana tracks. But "the feeling among automobilists that the Americans had transgressed the rules of the race somewhat when they employed horses to get through the snow and took advantage of railway roadbeds to get over bad places was lessened when it was learned the French and Italian crews are doing the same thing," the *Chicago Daily Tribune* observed.[130]

Jaded Leaders Reach Chicago

Late in the afternoon on February 25, 14 days from New York City, during a sleet and rain storm, the Thomas became the first New York–Paris racer to reach Chicago. (See Table 2.3.) In tribute, a brass band played "Love Me and the World Is Mine." Jack Banta of the Chicago Automobile Club drove the pilot car that led the Thomas Flyer into the city, the *Chicago Daily Tribune* reported:

> Other automobiles joined the procession, and all the way to South Chicago cheering men and boys lined the street. Just outside South Chicago a horse and wagon blocked the roadway. A collision appeared inevitable, but Banta turned into a snowdrift and plowed on. The horse jumped a fence.

> Though the drive downtown was made through a sleet storm, the enthusiasm of the crowds was comforting. In front of the automobile club a thousand persons pressed round the car and sought to shake hands with Roberts. He was forced to take off his head gear and a smile came to his face for the first time in days.[131]

The car "looked like the deck of a ship in a hurricane with the crew in drenched oilskins clinging to the sides." Though Roberts had weathered the storm, he reported losing 20 pounds in the previous two days of nearly around-the-clock shoveling and pushing. "The men, including Roberts, have lost all count of days, nights, and any events not connected with the race," Williams wrote in the *New York Times*. "The only words that can awaken signs of intelligence in them are 'Zust' and 'De Dion.' When they hear these names uttered in loud tones the jaded men start up from their chairs and shout wildly: 'Where? Where?'"[132]

Despite his triumphal entry, Roberts trailed one New York auto into Chicago. The Studebaker "war car" that left six days behind the New York–Paris racers arrived 20 minutes before the Thomas, according to the *Chicago Daily Tribune*. The dispatch-bearing Studebaker, however, "was worked continuously in relays by drivers who traveled ahead in trains," the newspaper noted. Comparing their running times, the Thomas had made better headway than the Studebaker, at least as far as Toledo, the *New York Times* contended.[133]

A race developed when the De Dion and Zust arrived together on February 26, a day behind the Thomas. (See Figs. 2.22 and 2.23.) Because the Zust waited $5\frac{1}{2}$ hours for the De Dion earlier in the day, De Dion driver St. Chaffray had agreed to let the Zust lead, according to Sirtori. But at an intersection near Chicago's Jackson Park, St. Chaffray suddenly shot past the Zust. Both drivers were soon racing for the Chicago Automobile Club "at top speed, reckless of the danger to themselves," St. Chaffray wrote in his daily *Times* dispatch.[134] Sirtori, however, swung past the De Dion on the last turn and pulled up first at the clubhouse.

Table 2.3 Daily Progress and Position of the Racers in the United States through February 26, 1908*

Day/Date	De Dion	Motobloc	Protos	Sizaire	Thomas	Zust
1 - 2/12	Hudson, N.Y.-3	Peekskill, N.Y.-5	Poughkeepsie, N.Y.-4	Crugar's Station, N.Y.-6	Hudson, N.Y.-1	Hudson, N.Y.-2
2 - 2/13	Amsterdam, N.Y.-3	Hudson, N.Y.-5	Albany, N.Y.-4	Crugar's Station, N.Y.-6	Fonda, N.Y.-2	Fonda, N.Y.-1
3 - 2/14	Canastota, N.Y.-2	Albany, N.Y.-5	Utica, N.Y.-4	Peekskill, N.Y.-6	Canastota, N.Y.-3	Canastota, N.Y.-1
4 - 2/15	Geneva, N.Y.-1	Wadsworth, N.Y.-5	Syracuse, N.Y.-4	Peekskill, N.Y.-6	Geneva, N.Y.-1	Geneva, N.Y.-1
5 - 2/16	Buffalo, N.Y.-2	Syracuse, N.Y.-5	Geneva, N.Y.-4	Red Hook, N.Y.-6	Buffalo, N.Y.-1	Rochester, N.Y.-3
6 - 2/17	Erie, Pa.-1	Geneva, N.Y.-5	Buffalo, N.Y.-4	Red Hook, N.Y.-6	Erie, Pa.-2	Ripley, N.Y.-3
7 - 2/18	Fremont, Ohio-2	Buffalo, N.Y.-5	Erie, Pa.-4	--withdrawn--	Toledo, Ohio-1	Cleveland-3
8 - 2/19	Corunna, Ind.-2	State Line, N.Y./Pa.-5	Conneaut, Ohio-4		E Kendallville, Ind.-1	Bryan, Ohio-3
9 - 2/20	Kendallville, Ind.-2	Ashtabula, Ohio-5	Cleveland-4		Wawaka, Ind.-1	Waterloo, Ind.-3
10 - 2/21	Kendallville, Ind.-2	Elyria, Ohio-5	Norwalk, Ohio-4		Goshen, Ind.-1	Kendallville, Ind.-3
11 - 2/22	Kendallville, Ind.-3	Toledo, Ohio-5	Delta, Ohio-4		New Carlisle, Ind.-1	Ligonier, Ind.-2
12 - 2/23	Goshen, Ind.-3	Bryan, Ohio-5	Edgerton, Ohio-4		Michigan City, Ind.-1	South Bend, Ind.-2
13 - 2/24	New Carlisle, Ind.-3	W Waterloo, Ind.-5	Corunna, Ind.-4		Burdick, Ind.-1	New Carlisle, Ind.-2
14 - 2/25	Michigan City, Ind.-3	Wawaka, Ind.-5	Ligonier, Ind.-4		Chicago-1	Michigan City, Ind.-2
15 - 2/26	Chicago-3	Goshen, Ind.-5	Elkhart, Ind.-4		Chicago-1	Chicago-2

* As of midnight each day. Because the racers often traveled all night, this represents each car's last known daily position, as reported in newspapers and in the crewmen's own accounts.

Sources: Articles and racers' dispatches in the *New York Times*; local newspapers; Hans Koeppen (Protos), *Im Auto um die Welt*; Antonio Scarfoglio (Zust), *Round the World in a Motor-Car*; and George Schuster (Thomas), *The Longest Auto Race*.

Figs. 2.22 (top) and 2.23 (bottom) The De Dion and Zust in Hobart, Indiana, on Feb. 26. Top photo, St. Chaffray is driving the De Dion. His front-seat passenger is an unidentified guide. In the rear, left to right, are Autran and Hansen. Note the car's tape-wrapped frame. Bottom photo, Sirtori mans the wheel of the Zust, joined by Haaga in the front seat. Ruland stands behind them while Scarfoglio, without furs, sits in the back. (LOC, both photos numbered 414 478)

Resting and Repairing

The racers reached Chicago in various states of disrepair. The Thomas, De Dion and Zust crews each took 2¾ days for repairs in Chicago. The Motobloc, which reached Chicago in fourth place on March 3, stayed nearly four days for repairs. The last-place Protos spent almost three days making repairs after its March 4 arrival.

The Motobloc crew reported only minor repairs between New York City and Chicago. The Frenchmen lost a half day when a broken gas line forced them to leave their car overnight in a barn near Utica. Godard's race against a train in north-central Ohio on Day 11 nearly ended in a tie. "Godard had seen the train coming and had hoped to cross in good time," the *New York Times* said. "Putting on full speed he made a dash for the other side, and with good luck got by, but the train shaved off the Pennsylvania license tag he had forgotten to remove."[135]

During a snowy night drive east of Cleveland, the Protos struck a "submerged" fence post and broke a wheel. The Germans slept at a farmhouse and lost much of the next day repairing the wood-spoked wheel. Soon thereafter, on Day 14, Koeppen traveled ahead to Chicago "in search of extra tires and parts which had been shipped ahead from New York."[136] The icy, rutted roads had taken their toll on tires—and the men. The two remaining crew members, Mass and Knape, both had to grasp the steering wheel to keep the heavy Protos on the road. In Elkart, Indiana, on Day 15 (see Fig. 2.24) the Protos crew made tire and engine repairs. Still, just seven miles later, they had a "long delay … due to tire trouble."[137] They changed tires at the Studebaker factory in South Bend on February 28. While being towed by horses on March 1, the Protos broke a steering part, stranding the crew for three days in Chesterton, Indiana, barely 50 miles from Chicago.

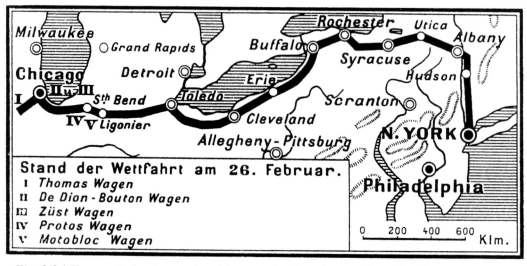

Fig. 2.24 Koeppen's book contains this map showing the position of each "wagen" as of Day 15, Feb. 26, 1908. (Im Auto um die Welt)

In Indiana, Schuster discovered "both of the housings carrying the transmission counter-shafts at the outside of the frame were breaking away." The Thomas factory misunderstood Schuster's telephoned instructions and sent ahead to Chicago a stock, rather than reinforced, pair of housings. Problems with these parts thus hampered the car as far as San Francisco.[138] The car hit a snowbank in Indiana hard enough to bend its steering rod, which was repaired in Chicago.

Foreigners File Protest

In Chicago, St. Chaffray's crew and racers on the Zust charged the leading Thomas with five rules violations. They concluded "we ... can no more consider the Thomas car on the same footing as the other contestants in the New York to Paris contest."[139] They alleged that the Americans:

1. made their Thomas into "virtually another car" during a visit to the E.R. Thomas Motor Company plant at Buffalo.
2. allowed horses to tow the Thomas, its engine turned off and radiator removed, for two days between South Bend and Michigan City.
3. were towed "by a trolley car over the tracks of the Interurban Street Railway Company" in "a certain section of Indiana."
4. drove part way between Michigan City and Chicago along railroad tracks, while the other cars used roads.
5. put their car on "sledges" (sleds) "at certain places."[140]

Company President E.R. Thomas personally denied charges that the auto was modified in Buffalo. Anticipating such a protest, the factory earlier had produced an affidavit signed by Factory Superintendent Frederick P. Nehrbas, who attested that workers installed an under-seat gas tank that wasn't finished in time to install before the start, replaced the rear bucket seats with a single bench seat and removed the car's fenders. Williams' account confirms this.

In his 1966 book, Schuster recalls that Thomas mechanics also installed heavier radius rods; replaced a bad cylinder; substituted a straight front axle for the curved one, which was scraping the snow; and put blocks above the axles to increase the car's road clearance. Mechanics replaced the metal fenders with pieces of heavy leather and relocated the oil reservoir so oil could flow by gravity, not pressure. Roberts added that factory workmen also removed "the prairie schooner [frame] after I had nearly been knocked senseless by being thrown against an upright."[141]

The protest was published on February 29, when Roberts crossed the Mississippi River at Clinton, Iowa, some 150 miles from Chicago. "From the time the Thomas car began to take a definite lead, the foreigners started their complaints about one thing or another," E.R. Thomas said. "I am not surprised at this latest outburst. They are fighting among themselves, while our people are going ahead steadily and surely. The whole truth of the matter seems to be that the foreigners in this race are bad losers."[142]

Race "Jars Peace of Nations"

"The protest of the French and Italians sounds to me like sour grapes," agreed Roberts, who denied the car was ever towed with its engine off. Further, he said that near Geneva, New York, on Day 4, the De Dion and Zust became the first cars to be towed by horses—a reference to their troubles in the Montezuma Swamp. But as conceded by Chicago Thomas agent C.A. Coey, who accompanied the racers through northwestern Indiana, "On different occasions it was necessary to stop the motor in order that the horses towing the car would not become frightened. This was at the request of the farmers."[143] He added: "It must not be forgotten that we broke the roads for the contestants following."[144]

While in Chicago, Hans Hansen announced that he had quit the De Dion. Before a notary public, he signed an affidavit saying the De Dion and Zust cars, like the Thomas, had been towed with their engines off. Moreover, Hansen charged that Paul Picard, a Chicago Frenchman who drove a pilot car in Indiana, knowingly directed the Thomas down a bad road from Michigan City but later guided the foreign racers down a better road. "Motor Race Jars Peace of Nations," a *Chicago Daily Tribune* headline blared. "Capt. Hans Hansen Charges Foreign Plot Against American Car." But these were only "dreaming stories," St. Chaffray scoffed.[145] Picard's response was that he was assigned to guide the De Dion and Zust only, and was not asked to provide the Americans with road information. Furthermore, he said he lent money to the Thomas crewmen so they could pay farmers who hauled them into Michigan City.

Roberts refuted the third charge leveled by the De Dion and Zust crews, saying an electric trolley ran ahead of the Thomas at Goshen to prevent accidents but never towed the car. Williams confirmed this. Concerning the fourth charge, Coey said driving along railroad tracks without flanged wheels was allowable under the rules "and all contestants had the same privilege, and did use the rails before reaching South Bend."[146] Roberts claimed the Americans always sought permission to use the rails. Contradicting Roberts, however, Williams wrote that the Americans had no time to seek permits to drive the tracks from Burdick to Hobart.

"At no time was the car on sleighs," Roberts contended, which was true, strictly speaking. But the *New York Times* and *Chicago Daily Tribune* both recount a short ride the Thomas took on a flat-skid "stone boat" early on Day 14, February 25, before reaching Chicago. The car spent 12 hours stuck in a deep snowdrift a frustrating 1½ miles from the rail station at Burdick, where the Thomas crew hoped to once again get on the railroad tracks. They finally bought a stone boat and struggled to load the auto on it. But the car slid off every time they hit a deep dip, Williams wrote. After gaining a half mile in two hours, Roberts abandoned the stone boat and had 12 horses pull the car the remaining miles, the *New York Times* said. According to the *Chicago Daily Tribune's* version, the sled broke after a mile. In yet a third account, Schuster recalled that the horses couldn't even budge the sled.

The Harry S. Houpt Company, co-sponsor of the Thomas entry, maintained that any violations were made "innocently on account of the lack of very definite rules."[147] If an investigation found violations, the Thomas would continue to Paris under the original rules regardless of the prize, the company said. As it was, race organizers found that the American entry had violated no rules. For the rest of the race across America, however, the leading foreign crews and the Thomas crew took turns charging one another with various acts of unsportsmanlike behavior.

Hansen Joins Thomas; Two Leave Protos

Hansen left the De Dion because, like Roberts, he quarreled with St. Chaffray. One such quarrel occurred at night while the De Dion lay stalled in Indiana four miles from the nearest town. Hansen and De Dion driver St. Chaffray "had a falling out … over the proper way to buck snowdrifts," as the *Chicago Daily Tribune* put it. "After the crisp atmosphere had cleared of the foreign gesturings it was found that each had challenged the other to the duello; that St. Chaffray had discharged Hansen, and the gallant captain had resigned."[148]

Added the *Omaha World-Herald*: "They could not agree, it seems, because St. Chaffray wished to drive in the fields while crossing snow covered Indiana and Captain Hansen wished to stick to the roads."[149] Meanwhile, Hansen had joined the Thomas crew. In Hansen's place, St. Chaffray hired Emanuel Lescares, a Chicago Automobile Club member, and upon reaching Omaha flayed Hansen in the local press:

> The captain never hunted for anyone up north, and I believe the Americans will get as tired of him as I did.
>
> He was not the man who went after Major Andrée, that was Dr. Nansen. He had me filled with his big talk, too, but I found him out, and you will see the dear captain will be left stranded in some of his Siberian villages with his Siberian friends he is always talking about, just as soon as the Americans find him out.
>
> Someone says it cost the Thomas company $8,000 to get Captain Hansen to go with the Americans. He would be dear at 8,000 cents.[150]

Hansen's actual pay was $100 per month, Dermot Cole writes in *Hard Driving: The 1908 Auto Race from New York to Paris*.[151] Also in Chicago, Knape and Maas, contending they were compelled to return to Germany, left the Protos car. The *Chicago Daily Tribune* later hinted at friction between the two men and Koeppen, who hired in their place O.W. Snyder.[152] A three-year German infantry veteran, Snyder was an engineer at the Woods Motor Vehicle Company, which made electric cars in Chicago, and "an expert chauffeur and mechanic." Said a Woods company official: "There was no dissension whatever in the German crew."[153] Oh, but there was, according to a story from Berlin:

> Information received here is to the effect that the men disagreed because the American press and public had been starring Lieut. Koeppen as the hero of the German expedition, whereas the men responsible for its conduct and the drivers of the car were Knape and Maas.

> The *New York Times's* correspondent has been requested by the Berlin *Zeitung [am] Mittag*, in whose name the Protos car is officially competing, to state that Lieut. Koeppen started on the trip exclusively as a passenger and correspondent. Knape and Maas, who are graduated engineers, as well as practical chauffeurs, wish general publicity to be given to this statement.[154]

Maas and Knape confirmed the story. The Berlin newspaper and the Protos company had each contributed $1,500 toward New York–Paris expenses; Koeppen, Knape and Maas invested $3,000 apiece and to this $12,000 total Koeppen added $2,000 "as a reserve fund," according to reports. "What troubled the two men [Knape and Maas] was … the realization that, while all were equal sharers in the expense, they alone were able to operate the car. Each is an expert driver and mechanician and, being amateurs, believed they were entitled to an equal glory." Koeppen, however, "sacrificed his entire personal fortune to complete this race," according to the *New York Times*. [155]

"Broke and Disgusted"

Other racers had fallen on financial hard times. In a February 29 telegram sent from New Carlisle, Indiana, to both the Chicago Automobile Club and *New York Times*, the Motobloc and Protos crews—still struggling toward Chicago—filed a protest against Indiana farmers:

> The peasants demand $3 per mile for helping us. Here at New Carlisle they charged $5 each to permit us to sleep on the ground. Peasants along the way have filled up [the] road dug by leading cars, so as to help the Thomas car. They make us hire as many horses as they please, so as to get as much money out of us as they can. We are broke and disgusted. Want you to intervene in our protection.[156]

In Indiana, Godard said, three 50-cent meals cost $5; $1 hotel rooms cost $5; renting one horse cost $5. "It is one strange country, that Indiana; everything you buy is $5." As Koeppen on the Protos put it: "The Indiana ways are not yet good."[157] Koeppen, however, later told the *Chicago Daily Tribune* that his only complaint was with an overcharging New Carlisle hotelkeeper. The *New York Times* reacted with a harsh editorial; the Chicago Automobile Club took more direct means, as "local cars were sent into Indiana to rescue them from the avaricious Hoosiers." More to the point, a messenger carried "a wad of money to the distressed autoists."[158]

The claim that Indiana farmers shoveled snow back onto the roads to impede racers resulted from "the natural irritability of tired and discouraged men," the *Times* editorialized.[159] But the newspaper and all crewmen felt the Indianans had taken advantage of the racers' helplessness. Paying $10 per mile for horses and $1 per mile for snow-shovelers, the Thomas crew spent $800 of company money to travel 64 miles from South Bend to Hobart. The deep drift near Burdick that ensnared the Thomas cost the American crew $95. The foreign crews faced similar expenses. "I have been thinking of buying those horses, since it would be much less dear to purchase them," said Godard of the Motobloc, who spent $120 to gain 30 miles. "Those Indiana people are like a pack of wolves."[160]

"It was estimated that the Indiana Hoosiers had benefited by over $2,500 besides having their roads opened by the racers," according to the *New York Times*. A half-mile-long procession of farmers in buggies, shut in for a week and hoping to get to church, followed the snow-shoveling Thomas crew near Rolling Prairie, Indiana, on Sunday, February 23. "The farmers do nothing," Roberts complained. "We hired their horses and paid them for the privilege of doing their work."[161]

Thieves Hit Motobloc

The Motobloc crew was especially sour. Thieves hit the Motobloc as it was locked in a barn at an overnight stop in Wawaka, Indiana, carrying away tools, spare parts, goggles, coats, gloves, Livier's movie camera and other loose items worth a total of $1,000. Godard claimed local police showed little interest in his plight. The Studebaker company in nearby South Bend invited the French motorists to use its tools to make needed repairs. The Chicago Automobile Club offered a $250 reward for the recovery of the items or capture of the thieves. Likewise, the AAA Racing Board offered $250 for the recovery of goods or the arrest and conviction of the thieves, partly "to show that we are anxious to give the foreign autoists a square deal in every respect during their trip through the country."[162] Godard personally offered $100 toward the reward. "Sheriff Stanley" eventually recovered the items March 7 in a house-to-house search of Wawaka, the *Chicago Daily Tribune* reported.[163]

Given the evidence, "we are under the painful necessity to believe that rustic greed has manifested itself in a way to bring the maximum of humiliation upon us all," said the *Times* editorial. "It is incredible ... that anybody in Indiana is so utterly stupid as to think that he can help the American competitor in the race by putting obstacles in the way of the foreigners or giving the latter anything less than the same assistance he received."[164]

But the contestants' good experiences outweighed the bad, even in Indiana, where some farmers kept lights burning all night to guide the racers. At Corunna, where the De Dion broke down, "two good, willing farmers opened to us their door and gave us beds," reported St. Chaffray. Later, in rural Illinois, "farmers came out of farmhouses and inquired if anything was needed," he said. Earlier, an elderly farmer blocked the Motobloc's path at East Pembroke, New York, forcing Godard to stop. The man, "who had his arms filled with apples, seized the opportunity to place them in the car, and, shaking hands with the party, wished them luck," the *New York Times* reported. "In France you find not that," Godard raved. "Ah, the Americans, they are fine!"[165]

Iowa Roads Ensnare Thomas

West of Chicago, the cars began to string out. The Thomas car left Chicago on February 28, Day 17, followed a day later by the De Dion and Zust, and managed to remain about 100 miles ahead through the Iowa mud to Omaha. Drier roads allowed the Thomas to speed across Nebraska in three full days. The Americans thereby reached Cheyenne, Wyoming, on Day 26, March 8, more than 500 miles ahead of the second-place Zust, stranded in Omaha for a day of repairs. (See Fig. 2.25.)

Iowa breakdowns had left the De Dion far behind the Zust. As Wyoming's mountain ranges slowed the leading Thomas, the Zust narrowed the gap between them. The Italians likewise gained slightly on the leaders when the Thomas broke its transmission in remote Nevada. Day 34, March 16, found the Thomas in western Utah, Zust in western Wyoming, De Dion in central Nebraska and Motobloc and Protos in western Iowa. (See Table 2.4.)

Some 3,000 Chicagoans turned out at 10:30 a.m. February 28 to cheer the departing American racers, Williams estimated in his *New York Times* account. Because of his crew's grueling all-night attempt to dig out of a snowdrift near Burdick, Roberts vowed to henceforth travel only by day. While in Chicago, Roberts bought three tasseled knit caps: red for himself, white for Schuster and blue for Mason B. Hatch, sales manager of the Thomas company. Hatch rode in a Thomas pilot car from Buffalo to Chicago but in the race car from Chicago to Boone, Iowa, press accounts reveal.

Roberts crossed the Mississippi River over the toll bridge at Clinton, Iowa, on February 29, Leap Year's Day. There, he found such deep frozen ruts that he occasionally detoured through farm fields. In the absence of road signs, the crew had to ask for directions, reporter Williams said. "The farmers along the route were all kind and meant well, but they knew less of directions or distances than the Hoosiers of Indiana." The Iowa farmers, however, knew what more rain would

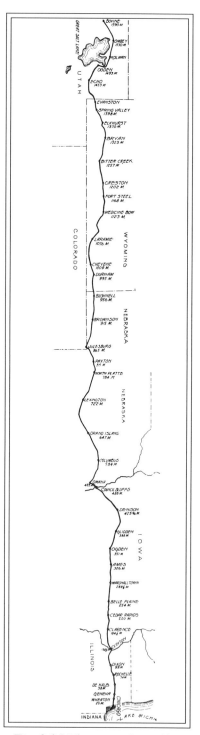

Fig. 2.25 The route from Chicago to northern Utah. (March 12, 1908, Motor Age/NAM)

Table 2.4 Daily Progress and Position of the Racers in the United States from February 27 to March 16, 1908[*]

Day/Date	De Dion	Motobloc	Protos	Thomas	Zust
16 - 2/27	Chicago-3	South Bend, Ind.-4	South Bend, Ind.-5	Chicago-1	Chicago-2
17 - 2/28	Chicago-3	New Carlisle, Ind.-4	New Carlisle, Ind.-4	Rochelle, Ill.-1	Chicago-2
18 - 2/29	Rochelle, Ill.-2	Rolling Prairie, Ind.-3	Rolling Prairie, Ind.-3	Clarence, Iowa-1	Rochelle, Ill.-2
19 - 3/1	Calamus, Iowa-2	Michigan City, Ind.-5	Chesterton, Ind.-4	Belle Plaine, Iowa-1	Calamus, Iowa-3
20 - 3/2	Cedar Rapids, Iowa-3	Chesterton, Ind.-5	Chesterton, Ind.-4	Ogden, Iowa-1	Cedar Rapids, Iowa-2
21 - 3/3	Cedar Rapids, Iowa-3	Chicago-4	Chesterton, Ind.-5	Logan, Iowa-1	Ames, Iowa-2
22 - 3/4	Cedar Rapids, Iowa-3	Chicago-4	Chicago-5	Omaha, Neb.-1	Vail, Iowa-2
23 - 3/5	Cedar Rapids, Iowa-3	Chicago-4	Chicago-5	Columbus, Neb.-1	Denison, Iowa-2
24 - 3/6	Cedar Rapids, Iowa-3	Chicago-4	Chicago-5	Lexington, Neb.-1	Woodbine, Iowa-2
25 - 3/7	Cedar Rapids, Iowa-3	Maple Park, Ill.-4	Geneva, Ill.-5	Sidney, Neb.-1	Omaha, Neb.-2
26 - 3/8	Le Grand, Iowa-3	DeKalb, Ill.-5	Rochelle, Ill.-4	Cheyenne, Wyo.-1	Omaha, Neb.-2
27 - 3/9	Marshalltown, Iowa-3	DeKalb, Ill.-5	Clinton, Iowa-4	Laramie, Wyo.-1	Grand Island, Neb.-2
28 - 3/10	Marshalltown, Iowa-3	Clinton, Iowa[1]-5	Cedar Rapids, Iowa-4	Walcott, Wyo.-1	Paxton, Neb.-2
29 - 3/11	Boone, Iowa-3	Clinton, Iowa-5	Cedar Rapids, Iowa-4	Bitter Creek, Wyo.-1	W Ogallala, Neb.-2
30 - 3/12	N Crescent, Iowa-3	Clinton, Iowa-5	Marshalltown, Iowa-4	Granger, Wyo.-1	Cheyenne, Wyo.-2
31 - 3/13	Crescent, Iowa-3	Cedar Rapids, Iowa-5	Ames, Iowa-4	Evanston, Wyo.-1	SE Medicine Bow, Wyo.-2
32 - 3/14	Crescent, Iowa-3	Tama, Iowa-5	Ames, Iowa-4	ca. Morgan, Utah-1	Dana, Wyo.-2
33 - 3/15	Crescent, Iowa-3	E Ames, Iowa-5	Ames, Iowa-4	Ogden, Utah-1	Rawlins, Wyo.-2
34 - 3/16	ca. Grand Island, Neb.-3	Ogden, Iowa-5	Denison, Iowa-4	Promontory, Utah-1	Rock Springs, Wyo.-2

[*] As of midnight each day. Because the racers often traveled all night, this represents each car's last known daily position, as reported in newspapers and in the crewmen's own accounts.

[1] Driver Charles Godard shipped the Motobloc by rail to Clinton for repairs. Afterward, he drove east into Illinois and back to Clinton to make up for the 80-mile railroad trip.

97

do to the steeply pitched, slippery roads, and urged Roberts to speed along. As it was, "Automobiling in Iowa at present is claimed by experts to be a delusion and a snare," the *Omaha World-Herald* observed.[166]

At Clarence, Iowa, local firefighters unrolled their hoses and blasted the mud off the Thomas. "This was so effective that we visited fire stations at several later stops," Schuster recalled.[167] Williams left the car that night at Clarence to return to the *New York Times*. The long travel days and one-hour time difference between the Midwest and East Coast had made it difficult for him to write and file his daily stories on time.

Females Flay Flag

Though he didn't care for their muddy roads, as Roberts wrote from eastern Iowa, "the Western people have won the first place in my heart.... When we find it necessary to cut a wire fence to cross a field to avoid a piece of bad road[,] instead of paying a big price for the small damage, we find the farmer beside the road with wire cutters in hand to do the work at no expense with a hearty 'Good luck to you, boys,' to speed us on our way."[168]

Crossing a stream between Clarence and Belle Plaine, Iowa, on March 1, a plank bridge gave way "and dropped the rear wheels of the car five feet and one of my companions ten feet to the Iowa mud," Roberts wrote in a dispatch from Belle Plaine.[169] Hatch, the unlucky companion who was tossed out of his seat, escaped unharmed, as did the automobile. The men reached Logan, Iowa, March 3, tired from frequent walks through mud up to two feet deep, evidently to push the car or lighten the auto on hills.

"It seems like a holiday in every small town we pass through," Roberts declared. "People have even decorated their houses with flags, and every school turns its pupils out to see the American car going by in the great New York to Paris race."[170] But many Iowans wanted more than just a look at the leading racer. Roberts recalled one such instance near Logan, where Miller and Schuster stopped the Thomas to clean mud from its radiator:

> The girls are always tearing pieces from the American flag we fly[;] in fact at Logan there was nothing left but the stars.... As I was sitting in the car while Schuster and Miller fussed with the radiator a girl grabbed the flag and ran. I chased her up the road through the mud and finally caught her. She gave up the flag with a laugh and gave me a flower she had in her hair. She told me to take it to Paris with the flag and when we left she threw me a kiss.[171]

Grand Reception at Omaha

When the Thomas car crossed the Missouri River in a drizzling rain just before noon on March 4, Omaha gave the Americans "the grandest reception we have had, not excepting New York City," Roberts said. Hatch, too, called it bigger than the sendoffs from New York City and Chicago. "When we reached the bridge between Council Bluffs and Omaha eight cannon were fired, and every whistle in the city was let loose," according to Roberts.[172] The din disrupted a murder trial. Police on horses and on foot were needed to clear a path to the Thomas garage through "mammoth crowds" lining the street for a mile. New crewman Hans Hansen, formerly with the De Dion, met the car outside Council Bluffs, Iowa. His first ride in the American car was into Omaha. (See Fig. 2.26.)

Roberts arrived in Nebraska cursing the "fields of mud" he had crossed in Iowa over the past 200 miles.[173] "It was with great satisfaction that they washed up and felt their feet once more on terra firma," quipped the *Omaha World-Herald*. A banquet at the Rome Hotel revived Roberts, who afterward accompanied Hansen to the City Auditorium's roller-skating rink. "A race was proposed and in the contest Roberts won handily." Added the *Omaha Daily Bee*: "It was a great sight to see the big Norwegian Captain Hansen try to apply ice methods to the rollers, but a few falls did not daunt him in the least."[174]

Fig. 2.26 Roberts, left, chats with new teammate Hans Hansen (wearing his Argentine naval uniform) in Omaha's Rome Hotel. (NAHC)

Mechanics washed the Thomas, fitted it with new tires and added "some leather aprons to keep the mud from the body of the car." Meanwhile, M.E. Smith & Company, an Omaha clothes maker, outfitted Hansen, Miller, Roberts and Schuster with free "corduroy trousers, flannel shirts, caps, gloves and, best of all, heavy sheepskin-lined overcoats," the *Omaha Daily Bee* said. The men agreed to mail them back from Paris in exchange for new outfits. "When they pulled out of the garage they were bundled almost beyond recognition."[175]

After a 23-hour stay, the Thomas left Omaha at 10:30 a.m. March 5, cheered by "at least 10,000 people," Roberts estimated.[176] A new passenger was aboard, at least at intervals: *New York Times* photographer Charlie Duprez. Omaha's paved terra firma led directly into 20 miles of "yellow clay, soft and sticky," according to Omaha Thomas agent H.E. Fredrickson. "Going up some of the hills we had to get out and push," said Fredrickson, who assisted Dr. H. Nelson Jackson and Sewall K. Crocker, the first transcontinental drivers, when they stopped in Omaha in 1903.[177] In his own Thomas car, Fredrickson would accompany Roberts to Cheyenne.

"Iron Driver" Bows Out

But the roads began improving the next day. One section between Columbus and Lexington, Nebraska, was hard sand with grass growing in the middle. And then, "there was no road," Roberts recalled. "We ran across the open prairie on the grass, following the railroad."[178] More than ever, crowds lionized the American crew. "It has got so now that wherever we stop overnight we have to put a rope around the car and let every one in the town pass by," Roberts wrote from Lexington. Residents of the western Nebraska town, to his confessed pleasure, freely called him by his new newspaper nicknames of "Get-there-Roberts" and "Iron Driver."[179] Between Lexington and Sidney, Nebraska, farmers telephoned their neighbors with news of the leader's progress "so that all were down at the gates to meet us and wave the American flag at us," wrote Fredrickson. "This race has done more to interest them in automobiles than all the [automobile] shows."[180]

On the desolate stretch from Sidney to Cheyenne on March 8, "we saw only ten houses the entire distance, and passed one prairie schooner," Roberts said.[181] A group of cowboys greeted the race leaders and its automobile escorts outside Cheyenne, as the *New York Times* recounted. "They were mounted on spirited mustangs that were thoroughly frightened by the cars, and cut up all sorts of capers. The yelling riders wheeled in ahead of the procession and tore through the streets in mad fashion. It was a most picturesque demonstration, and Roberts and his party enjoyed it hugely." (See Fig. 2.27.)

As planned, at Cheyenne Roberts turned over "Old Baby"[182]—his nickname for the car—to E. Linn Mathewson, 23. With his father, Mathewson ran a Denver auto agency that in 1908 was selling Columbia, Oldsmobile and Thomas autos, as well as Columbus and Woods electrics. Five years previously while working for an Oldsmobile dealer, the younger Mathewson persuaded Olds transcontinentalists Lester L. Whitman and Eugene I. Hammond to detour to his

Fig. 2.27 Packing the streets outside the Capitol Garage, a large crowd welcomes the Thomas to Cheyenne. Roberts stands in the car among a sea of hats, clutching a bouquet of flowers. Behind him, Miller receives his accolades sitting down. Schuster is not visible. (WDCR)

Denver agency for engine improvements. From Cheyenne, Roberts returned East to partici- pate in the Briarcliff Trophy road race—10 laps around a 32.4-mile route in Westchester County, north of New York City. Roberts' Thomas was still running and in 15th place among 22 starters when the winners were declared and the race was called off after eight laps.[183]

Tireless Mathewson Heads West

In his own five-year racing career, Mathewson "has won important races on the Western tracks and on the road, and has a reputation of being tireless in a machine." When Mathewson, in an Oldsmobile, broke the Denver–Colorado Springs speed record in 1907, the *Rocky Moun- tain News* heralded the event with a large photo of the speed demon in action and the head- line—"Fastest Express Train Outsped By Motor Car."[184]

Steered by Mathewson and wearing two new tires, the Thomas left Cheyenne for Laramie the morning after its arrival. But Schuster, who demanded to take the wheel at Cheyenne for the rest of the race, staged a "one-man strike," delaying the scheduled 8 a.m. departure, the *Cheyenne Daily Leader* reported. At 10:30 a.m., the car emerged from the Capitol Garage to the cheers of 1,000 impatient spectators, and drove to the Inter Ocean Hotel.

> But at the hotel the car came to a standstill and there it remained until 11:25, while won- der at the delay and speculation as to its cause increased.

Meanwhile the management of the car had been attempting to get the recalcitrant Schuster back into line, but without effect. As the car stopped at the hotel matters were made worse by Schuster suddenly disappearing. Then the management of the car was utterly at sea…

So, while Captain Hans Hanson [sic], the Norwegian explorer and a member of the crew of the American car, kept the crowd in a good humor by alternating between cracking jokes in broken English and paying elaborate court to Miss Katherine Mackenzie, the Cheyenne beauty, the management of the big racer was in a decidedly uncomfortable predicament.[185]

Back in line at last, Schuster "resum[ed] his place in the mechanician's seat," and the car set out, joined for the 19-mile drive to Granite Cañon by an extra passenger—Miss Katherine Mackenzie. At Laramie, Schuster worked all night to replace another set of cracked counter-shaft housings. On March 10, Day 28, Mathewson and the Thomas crew left Laramie for Walcott, hitting 60 mph for the first 30 miles.

"While we ate lunch the women of Medicine Bow tied their handkerchiefs to [the Thomas] in the fond but foolish hope that the bits of linen would some day reach Paris," Mathewson wrote. "Certain members of the Thomas crew being sadly in need of handkerchiefs— enough said." Mathewson crossed the Medicine Bow River on two feet of ice and at Hanna approached the Medicine Bow Range of the Rocky Mountains. "With a terrific hill to face, I decided to ship all excess baggage onto Rawlins," he said. "We learn that the Italians are making good time through Nebraska. Wait until the Marconi fleet arrives in the Rocky Mountains, and we shall see whether their cathedral on stilts is any good on a side hill."[186]

The Thomas crew shoveled snow often on March 10 and the next day, until, west of Rawlins, they drove out of drifts into slush, mud, and, finally, dirt in crossing the Red Desert to Bitter Creek. Fording a stream between Bitter Creek and Granger on March 12, the car "took in so much water through the radiator that the fan was badly bent," wrote Mathewson, nursing a slightly sprained wrist after four days of driving. Farther west, driving between Rock Springs and Green River "was like pulling up the side of a house plastered with chewing gum."[187]

The critics of the Great Race had fallen silent, *Motor Age* noted on March 12, exactly one month after the start from Times Square: "The newspaper paragraphers have exhausted their witticisms and as the leader breaks his way toward the Pacific coast his progress is being followed by hundreds of thousands of readers of the press reports and the affair does not seem half so nonsensical as when it first was broached."[188]

Cruel, Hard Driving

To avoid 10 feet of snow in the hills, the Thomas crew on March 13 traveled the Union Pacific's tracks between Carter and Evanston, a distance Schuster estimated as 45 miles.

With a conductor aboard, the car ran as a "special train." It traveled with its left tires inside the rails and its right tires on the tie ends, striking many plates and bolts. Just a few miles after leaving Carter, both right tires popped within a mile of one another, Schuster said. Repairs took 30 minutes apiece. A third tire exploded near the entrance to the 5,900-foot-long Aspen Tunnel in southwestern Wyoming, through which the Thomas avoided miles of snowbound mountain roads. They arrived at 2 a.m. in Evanston as a four-hour trip stretched to 10 hours.[189] Besides ruining three tires, such "cruel, hard traveling" along the ties forced mechanic George Miller to work all night on other unspecified repairs, Mathewson wrote.[190]

Striking off by road the next morning, the Thomas dropped into a "snow sink" just two miles out of Evanston. While everyone else spent the next five hours digging out, their local guide, Evanston lawyer Payson W. Spaulding, walked back to Evanston to get permission to ride the rails for nine miles to reach Wasatch, Utah.

At Wasatch, the American crew got permission to remain on the rails for another nine miles to Castle Rock, where the Thomas again ventured onto the highway. The crewmen drove the rest of the day and all night, and slept in the open car—they'd discarded the top at Buffalo—during a rainstorm near Morgan while their pilot car went on to buy gas. At 9:30 a.m. March 15, Mathewson reached Ogden, where he would leave the car.

After a day spent installing another new set of countershaft housings and making other repairs in Ogden, Harold S. Brinker, 22, took the wheel. Mathewson and Brinker would meet again several weeks later in a 320-mile road race near Denver, staged on "the roughest road ever traveled by any set of cars" in a U.S. endurance race, according to *Motor Age*. This time, the New York–Paris teammates were competing against one another. Driving a 40-horsepower Thomas 40, Mathewson won the Rocky Mountain Endurance Run—an event that saw mechanical breakdowns sideline five of the eight starters. Suffering a broken water pipe, Brinker's Denver-made Colburn 40 was the first car to drop out.[191]

For the previous three years Brinker had acted as "the racing man of the San Francisco branch of the Thomas company."[192] As such, he made a San Francisco–Chicago run during which he conquered the Sierra Nevada, according to the *New York Times*. Brinker, who, like Mathewson, at times held the Denver–Colorado Springs speed record, was back in his native Denver by 1909, at least temporarily—employed as an auto agent and part-time racer. He soon moved to Cheyenne, where he sold and raced automobiles and built and flew airplanes, before his return to Denver. (See Fig. 2.28.)

In fact, Brinker may have continued to reside in Denver while racing in California during part of each year. He is listed in the 1908, 1909 and 1910 Denver city directories. By contrast, the San Francisco city directories for 1907, 1908 and 1909 do not name Brinker. In a long, colorful racing career, Brinker participated in contests that included the 1910 Galveston (Texas) Beach Races and a 1916 *Denver News-Times'* 132-mile road race between Denver and Laramie, Wyoming. Brinker's Cadillac beat all 21 competitors and lowered the speed record between the cities by 33 minutes.[193] He was "a splendid driver," Schuster recalled.[194]

Fig. 2.28 Brinker at the controls of an early wood-and-fabric biplane. (WDCR)

Doctor's Car(e) Heals Break

When UP officials refused to let the car use its 40-mile-long trestle across the northwestern part of the Great Salt Lake, the Thomas crew instead headed north around the lake through Promontory. Brinker drove through several sandstorms to reach Cobre, Nevada, March 17. (See Fig. 2.29.) He left Ely (see Fig. 2.30) late Wednesday afternoon, March 18, hoping to make the 30 miles to Preston, but Schuster recalls that the car got stuck in a bog southwest of Ely. The crew spent the night at the nearby "Veteran mining camp," according to Brinker. He hoped to reach Goldfield, Nevada, on March 20, travel 500 miles during the next two days and then sprint the last 400 miles to San Francisco in a day. "I have made the run from Los Angeles to Frisco in twenty-three hours several times, and as the roads are now in good condition I think I can improve upon that."[195]

But at 6 p.m. on March 19, while fording a river with steep, muddy banks near Twin Springs, the Thomas broke six teeth from its transmission drive pinion and cracked the transmission case. (See Fig. 2.31.) It was evening, but Schuster found and rented a horse for $20 and began riding west. He met eventually with a rescue car sent from Tonopah, 75 miles away, where Schuster persuaded a doctor to surrender the necessary parts from his Thomas. Schuster returned at noon on March 20, the crew installed the new parts by 8 p.m. and by 11:15 p.m. the car had reached Tonopah for the night, Brinker said.

Fig. 2.29 Moving cautiously to avoid damaging the machinery, Brinker crosses a set of railroad tracks near Cobre, Nevada. (AAMA)

Fig. 2.30 The Thomas as it appeared in Ely, Nevada, sporting its distinctive and useful skid planks, one per side. The crewman adjusting the load on the back of the car is unidentified. (NHS)

Fig. 2.31 Brinker eases the Thomas across a river at Twin Springs, Nevada, moments before the transmission gives way. (NAM)

By 9 a.m. March 21, Day 39, the Americans were off again, crossing into California that afternoon. Brinker recalls driving all night through Death Valley, which was "a mass of moving sand, and was strenuous work for the car."[196] Even with the tires partially deflated for better traction, everyone but Brinker had to push the car almost constantly. This and the frequent stops forced by the steaming radiator held the car's average speed to 2 mph on the eight-mile stretch from Stovepipe Wells west across Death Valley.

Cheering Throngs Greet Thomas

Onward Brinker drove throughout the following day, avoiding Daggett, California, and saving 158 miles by taking a more direct but slower northern route through the mountains. Expected in Daggett, the Thomas crew instead surprised residents of Mojave on March 22 by stopping for dinner at 5 p.m. Brinker pressed on to reach Bakersfield at 1 a.m. March 23, thus ending 388 miles of nearly continuous driving over 40 hours. The Thomas company was hoping to hold for 24 hours the steamer that was due to leave Seattle for Valdez on March 24. But the Thomas Flyer's unspecified "small breakdown at Selma" delayed its arrival at Fresno on March 23, preventing the Americans from making their rail connections at San Francisco.

Upon reaching Oakland on Tuesday afternoon, March 24, "the main street was so crowded that the Thomas car could proceed only with the greatest difficulty. More than once it was compelled to stop, while hundreds of enthusiasts blocked its path and insisted upon shaking hands with the American crew." The ferry crossed the bay to land them in San Francisco, where "the streets for blocks from the ferry up Market Street were absolutely jammed by cheering throngs. The scores of automobiles that followed every move of the racers kept their horns blowing and whistles in factories made a din."[197] (See Fig. 2.32.) Ending in San Francisco at 4:40 p.m., the Thomas Flyer's wintertime crossing had consumed 41 days, 8 hours, 25 minutes. Exactly how far it traveled between coasts is pure speculation. The *New York Times* said 3,832 miles, later revising it slightly to 3,836. The newspaper, however, computed its distances based on railroad mileage, "which is direct and totally different from the road the car follows," complained Roberts. "A fifty-mile stretch of track often means eighty miles of road, but the extra distance does not show in the records."[198] The Thomas company embraced the *Times'* figure, however. Thus for 3,836 miles and its elapsed time across the continent, the Thomas averaged 3.87 mph and 93 miles per day.

Fig. 2.32 Brinker steers the first-place Thomas through a San Francisco crowd. Miller, looking backward, is in the front seat. Behind him are Hansen, left, and Schuster.
(AAMA)

The crew put its car on public display at the Thomas agency in San Francisco, where Schuster, Miller, Hansen and others began two days of repairs. They included replacing the counter-shaft housings for the fourth time; bracing the frame to ease the strain on the housings; installing new driving chains; and replacing the transmission, wheels and badly sagging springs, Schuster said. (See Table 2.5.) Detractors said winter was waning in Alaska. Nonetheless, the American crew loaded its car on a steamer for Seattle and on to Valdez, arriving in Alaska on April 8 to continue the race to Nome.

Table 2.5 The Thomas Flyer's U.S. Breakdowns and Repairs in 1908 Race*

Day 1, Feb. 12	Replace spark plug and grind valves in Cylinder No. 4. (Schuster)
Day 2, Feb. 13	Snowdrifts bend the radius rods several times. Mechanic George Schuster straightens them with a jack, as he does repeatedly during ensuing three days. (Schuster)
Day 3, Feb. 14	Stop en route to Canastota, N.Y., "to overhaul the [drive] chains." (NYT)
Day 5, Feb. 16	Buffalo, N.Y.: Install one new cylinder and stronger radius rods. (NYT and Schuster)
Day 13, Feb. 24	Car bends "steering rod" in Indiana snowdrift. (NYT)
Days 14-16, Feb. 25-27	Chicago repairs: Straighten steering rod, replace cracked countershaft housings (a driveline part), tune "poorly" running engine. (NYT and Schuster)
Day 27, March 9	Laramie, Wyo.: Install second pair of countershaft housings. (Schuster)
Day 30, March 12	Granger, Wyo.: Unspecified "overhauling." (Mathewson)
Days 31-32, March 13-14	Evanston, Wyo.: Mechanic George Miller works all night repairing unspecified damage from running on railroad tracks. (Mathewson)
Day 35, March 15	Ogden, Utah: Install third pair of countershaft housings, perform unstated "other repairs." (Schuster)
Day 37, March 19	Twin Springs, Nev.: Replace transmission case and broken transmission gear. (Schuster)
Day 41, March 23	Selma, Calif.: Unspecified "small breakdown" delays car. (NYT)
San Francisco repairs	Replace countershaft housings for the fourth time; install a frame brace to reduce the stress on these housings; replace the transmission, springs, wheels and drive chains. (Schuster)

* Includes all reported repairs (see "sources") but excludes daily maintenance and adjustments.

Sources: (Mathewson) = E. Linn Mathewson's *New York Times* dispatches
(NYT) = *New York Times*
(Schuster) = George Schuster and Tom Mahoney, *The Longest Auto Race*

Zust, De Dion in "Glutinous Morass"

Still piloted by the Gearless car from Rochester, the Zust left Chicago February 29 with the De Dion, a day behind the Americans. The Illinois roads west of Chicago were "good for skating," said St. Chaffray. It was so icy he slipped and fell on the steps of his hotel in Rochelle, Illinois. "We crossed corn fields and found them better than the roads."[199] One morning in Iowa—apparently at Calamus—the Zust crew went out to discover two flat tires.

"Two long nails had been stuck in them—in the sides, mark you, where the tyres did not come into contact with the earth," Scarfoglio wrote.[200]

The De Dion broke a front spring and damaged its steering gear near Lowden, Iowa, on March 2, the second day of its 15-day struggle through Iowa mud. The crew jammed blocks of wood between the spring and frame to limp into Cedar Rapids, where spring repairs and attempts to brace a sagging frame occupied the next five full days. (See Fig. 2.33.) "The motor of the de Dion is a wonder," Lescares wrote in a letter to *Motor Age*, "but the body is too heavy for the chassis."[201] Breakdowns—some caused by sabotage, St. Chaffray charged—idled the De Dion for eight days in Iowa. The car remained a distant third all the way to San Francisco after such delays as turning upside-down in a Wyoming mudhole and bogging down in the desert sands of California's Death Valley.

Thousands greeted the De Dion and Zust on March 2, when they reached Cedar Rapids at mid-afternoon, the Zust 45 minutes in the lead, according to the *Cedar Rapids Evening Gazette*. A band waited as the Zust, now in second place ahead of the De Dion, arrived at Jefferson in east-central Iowa on March 4. Pulling into Jefferson, the Zust suddenly tilted sharply to the right; sympathetic spectators rushed toward the crew, Scarfoglio related:

Fig. 2.33 The De Dion arrives in a Cedar Rapids garage, its frame sagging visibly. Earlier in the trip, crewmen lightened the car by cutting down its cargo box. St. Chaffray is in the front seat, foreground; Autran is at the wheel; and Lescares holds a Chicago Motor Club pennant. (March 12, 1908, Motor Age)

For a moment we could not understand what had happened.... The bearing of one of the back wheels had split, and the wheel, detached from the axle, was taking flight across the fields. By a miracle the axle itself was not broken. We are assisted by a body of ladies, who manage spanner and hammer with skilful [sic] hands, and do not disdain to cover themselves with grime and oil in their anxiety to help us.[202]

The Italians left Vail, Iowa, March 5 for the easy half-day's drive to Omaha. But because recent rains had turned the roads to "a glutinous morass," Scarfoglio said, the 75-mile trip took three days. "As in the snow, Haaga, with great india-rubber boots up to his thighs, preceded us, taking soundings. I and Ruland—an American who is accompanying us as far as San Francisco—followed, and the car travelled as it could, sliding and labouring, towards Denison." They advanced nine miles for the day.[203]

Spring Cracks on Tracks

On March 6, the Illinois Central Railroad ran the Zust as a special "Second No. 1" train, following the Chicago-Omaha Limited over the rails from Denison. "First No. 1, the limited, made fifty miles an hour. 'Second No. 1' made just four miles during the first hour," the *New York Times* recounted.[204] The crew had to lay planks between widely spaced ties so the Zust could cross the Boyer River bridge near Arion. The car broke two leaves in a rear spring when the wheels hung up on a frog, or switch, during the crossing. The crew braced the spring with a block of wood. It took an hour to cross this bridge, and there were several others ahead. The Zust drove 30 miles to Woodbine by rail on March 6 and 45 miles into Omaha—most of it by rail—the next day.

A crowd jammed the Illinois Central station at Council Bluffs, Iowa, where the Zust arrived after 8 p.m. March 7. From Council Bluffs, the Italians received an auto escort over the Missouri River bridge into Omaha. "Give me good roads and I will beat the American car," Sirtori declared in Omaha. "However, since the American has violated all the rules of the race and will probably be put out, we consider ourselves first."[205] According to the *Omaha Daily News*,

> A platoon of police had to surround the foreign car at Council Bluffs to keep souvenir hunters from carrying it away in pieces. The Italians have lost two robes and one leather top since they left New York, all literally cut to pieces by persons wanting souvenirs....

> "American girls, they will take anything they can find loose," said Sartori [sic], who is mourning the loss of a small Italian flag, "but they are so nice and most of them so pretty, we do not get very angry."[206]

Repairs consumed that night, the next day and part of the following morning. Machinists at the Union Pacific shops worked all day and night Sunday, March 8, making two new spring leaves. An Omaha silversmith boasted of repairing the Zust's radiator, which had leaked since the start of the race. (See Fig. 2.34.) The Italians used equipment at the R.R. Kimball Company garage "to take their machine entirely apart and Mechanician Haaga scattered the automobile all over the place while he thoroughly cleaned it," the *Omaha Daily News* reported. "All day Sunday the front of the garage was besieged with people, who were admitted, a few at a time, to see the car." Because the Italians shipped ahead "a considerable amount of extra weight," it was a lighter Zust that left Omaha at 10 a.m. Monday, March 9, for the open prairie.[207]

Italian Racing Car

"THE ZÜST Stops"

IN OMAHA long enough to have
its radiator breakage repaired
and refinished by the

OMAHA SILVER CO., Inc.

KEMPER, HEMPHILL & BUCKINGHAM

SILVERSMITHS

Phone Doug. 1773 All Kinds Plating

AUTO LAMPS and RADIATORS Repaired and Refinished ...,

No Matter How Badly Damaged

Between Farnam and Harney 314 S. 13th

Fig. 2.34 The Zust's visit was advertising fodder for the Omaha Silver Company. (March 15, 1908, Omaha Daily Bee*)*

Zust Founders on Sea of Grass

"The eye is almost hypnotised by this colossal sea of grass upon which there is not a house or a tree," Scarfoglio wrote. "Each day's journey is like a great parenthesis across which the brain of each of us rides alone. We have one single point of contact, one common sentiment, one common need—to push forward."[208]

Two days out of Omaha and two-thirds of the way across Nebraska, the Zust stopped in Paxton late on March 10 "because a terrible leap into a ditch had broken a pinion in the car and almost smashed our ribs," Scarfoglio wrote. (See Fig. 2.35.) The broken part was actually the sprocket in the chain-drive mechanism, said Ruland. The UP delayed a train 15 minutes in Cheyenne waiting for the correct replacement part. It arrived in Paxton late in the afternoon on March 11; Haaga installed it immediately. While making repairs, the Italians discovered a crack in the right side of the frame between the springs. "Only a thin tongue of steel held it together," said Scarfoglio. They couldn't repair it in Paxton so they drove all night to Cheyenne, where the car was "thoroughly overhauled at the Union Pacific shops."[209] (See Fig. 2.36.)

The Italians encountered a series of scares and vexing delays west of Cheyenne, some of them "due to their persistence in running over rough country in the dark," as the *New York Times* put it.[210] West of Cheyenne en route to Medicine Bow at night, the Zust left the road and stuck in the mud. Someone walked into Medicine Bow for horses and the car finally reached town at 4 a.m. At Fort Steele, the Thomas had crossed the frozen North Platte River on the ice but the

Fig. 2.35 The Zust at the Paxton, Nebraska, blacksmith shop where it sought repairs.
(Round the World in a Motor-Car)

Fig. 2.36 Bearing Italian and U.S. flags, a Reo auto escorts the Zust (second in line) into Cheyenne. Sirtori is driving with Haaga by his side. Behind them, left to right, are Ruland and Scarfoglio. (WDCR)

Italians spent several hours wrestling their heavier Zust up and down steep embankments to use the UP railroad bridge.

As the Zust raced toward Granger at or near nightfall on March 17, a deep hole suddenly appeared in the road ahead. Sirtori "did not see the yawning chasm until he was right upon it, and then applied the brakes and reversed the engine in an effort to avoid plunging over the brink … It seemed as though the four men must be hurled to death in another instant. The Italian stuck to his seat, however, even when the front wheels sank over the edge," the *New York Times* recounted. "The axle, however, caught on the brink, and the car hung an instant and then stopped, while the four men got out and lifted it out of its precarious position." Scarfoglio recalls that Sirtori pushed him from the car as it hurtled toward the chasm, shouting, "Quick, quick! Save yourself!"[211] Though dramatic, Scarfoglio's description is suspect, for Sirtori drove from the right side of the front seat while Scarfoglio rode on the left side of the back seat. Freeing the car took two hours; the crew was further delayed finding a detour.

A railroad had dug the pit for railway fill, said Ruland. He described the episode as just one of four instances when the crew was "in imminent danger of death" between New York City and Ogden, where he left the car. Two others also occurred in Wyoming. Once, a mountain trail began to collapse toward a 500-foot cliff. "The whole road slid absolutely to the edge of the precipice and lodged there against a boulder." It happened so quickly that no one had time to jump.

Another time, a bearing broke and the Zust lost a wheel, as it did in Jefferson, Iowa. This time, the car was traveling at 30 mph. When the axle end hit the ground, "we were hurled out on the turf. Fortunately no one was injured … though how we escaped I cannot now understand." The fourth death-defying experience came in Iowa along the fill of the Illinois Central tracks, which skirted many "great sink holes." Near one of these, the crust of the road gave way, dropping the Zust over an embankment. Ruland was amazed the car didn't turn over and kill them all.[212]

Railroad Helps, Hinders

In western Wyoming (see Fig. 2.37) the Italian crew found "a most remarkable guide, a big, ruddy man dressed in goat skins, the very type of the coarse, wild cowboy. A lord in this limitless land, the possessor of innumerable flocks and herds," wrote an impressed

Fig. 2.37 Haaga crouches to check the clearance as Sirtori eases the Zust into "a dried water-course" at an undisclosed location—most likely Wyoming. The unidentified third man is probably a guide. Scarfoglio is behind the camera. (Round the World in a Motor-Car)

Scarfoglio.[213] After a banquet for the Zust crew that night in Rock Springs, however, their drunken guide began shooting at the electric lights around town. Though he spent the night in jail, the wild cowboy was ready to continue with the Zust the next morning, according to Scarfoglio.

Zust race organizers set up tire relay points across the West. By the time the car reached Granger, it had used 10 complete sets of tires and many extra inner tubes, Ruland said. Just west of Granger on March 18, the Zust dropped off a UP railroad bridge into a muddy stream, where it remained for the next 30 hours. Hired horses balked at crossing the railroad trestle and couldn't swim the swollen stream that it bridged. On March 19, the railroad station master in Granger sent out "an engine and three or four trucks loaded with Japanese labourers and railway men, with poles, beams, jacks, and chains.... Holding on to the ropes and chains, the 150 Japanese gave one pull and the car came out of the mud like a hand out of a glove," Scarfoglio recounted.[214]

While traveling at night, the Zust crew ran out of gas at Spring Valley in the mountains of southwestern Wyoming, and had to wait for a supply to arrive from Evanston. Sirtori complained bitterly about being denied permission to use the Union Pacific's Aspen Tunnel. But by running with one wheel outside the rails, the Thomas had "simply scattered the ballast [rock fill] in all directions, and made it necessary to send section gangs all along the road and fix the ballast again," according to Ruland on the Zust. So "their refusal was wholly justified from their viewpoint, though we felt they should not have extended the courtesy to one car and not to another.... Luckily the refusal of the right to use the Wyoming tunnel did not delay us."[215]

If a foreign car had been the first one over the tracks, the Union Pacific replied, the railroad would have denied access to the trailing autos—including the Thomas—to prevent further damage. Moreover,

> We have rendered every facility in our power to all the racers. When repairs were necessary and outside shops could not do the work, the Union Pacific threw open its shops and placed every piece of machinery and every machine and machinist at their disposal. We even had our shops open on Sunday for the express purpose of assisting one of the foreign cars. In fact, but for the aid we rendered the French and Italian cars, they would probably not have gotten past Omaha at all.[216]

Broken Frame Dogs Zust

A *New York Times* headline on a wire report told of another adventure that supposedly befell the Zust crew near Spring Valley: "Italians on Zust Kill 20 Wolves." In California, Scarfoglio met "a man badly dressed, chewing tobacco, and with dirty finger-nails"—a Los Angeles newspaper reporter who had either concocted the myth of the killer wolves or was seeking to perpetuate another reporter's creation. As Scarfoglio relates, "in Spring

Valley the only victim of our guns was an innocent sparrow which was airing itself on a telegraph pole."[217]

At Ogden, where Ruland left the crew to return to New York City, the Zust received a new front axle and other repairs. In preparation for the desert crossings ahead, the crew reportedly modified the Zust by replacing the back seat with a large gas tank. This would preclude sleeping in the car, according to Scarfoglio, who neglected to reveal where the former occupants of the back seat would sit west of Ogden. "Hence the car was loaded with food, filled with water like a cistern, and we were given ... a tent, a quantity of ammunition, maps, and much advice."[218] The Italians left at noon March 22 to cover 80 miles to Kelton during a 12-hour driving day.

Scarfoglio believed someone sabotaged the car at Ogden. Northwest of the city, Haaga discovered that the source of a differential noise "was a fine fat nail, and it could not have found its way thither merely by mistake." On March 23, muddy ground forced the car to leave Kelton over the abandoned Southern Pacific railroad line. Because rain had washed out the fill rock between ties, Scarfoglio said, "it was like driving across a ploughed field." At Cheyenne, the Italians had fastened a strip of steel over the chassis crack they discovered in Nebraska. But pounding along the abandoned tracks had broken the chassis in front of that patch. The Southern Pacific, however, "extended a most unusual courtesy to the Italian crew" by sending a special locomotive and car out to carry the crippled Zust by rail back to Ogden.[219] There, despite Sirtori's earlier denunciation of the railroad, the UP again opened its shops to the racers, and shipped their car back to Kelton to resume its journey March 25.

Ruland was kinder than Sirtori. Afterward, he thanked the railroad for the use of its shops at Omaha and Ogden, for stopping a train at Paxton to unload repair parts, and for extricating the car near Spring Valley, Wyoming. Nonetheless, the crew took many chances in crossing railroad bridges over streams, he said. "We asked permission and were refused, so we simply took our lives in our hands and went over the rails any way."[220]

Zust at Finish: "Nothing is Right"

Forewarned, the Italians equipped themselves with "smoked glasses" but nonetheless found the light dazzling during their March 30 daytime crossing of Death Valley. "The heat became atrocious, insupportable," Scarfoglio wrote. On the soft sand, Haaga laid down canvas strips for Sirtori to drive over. But by early afternoon the car had covered only two miles in $3\frac{1}{2}$ hours. "At four o'clock we find three human skeletons. They were quite close together, about six or seven yards apart. The bones are as white as snow."[221]

Sirtori tried to follow Brinker's northern route from Death Valley to Mojave, California, but strayed too far south to land in Daggett. Approaching Los Angeles the next day, "the big car struck a bump in the road and burst the gasoline tank," but made it into the city for repairs.

Los Angeles "went fairly wild over the Italians," the *New York Times* said. "Even the best posted citizens had no idea there were so many Italians living here."[222] After a banquet that night, the crew drove on, even though the Zust had again broken its frame, Scarfoglio wrote.

The car arrived in San Francisco at 10:35 a.m. April 4, too late to catch a steamer that would have sped its arrival in Valdez, Alaska. "The American people on the Pacific Coast are killing my boys with kindness," complained Zust representative R.W. Vollmoeller. "The reports of enthusiastic welcomes and receptions are very gratifying to our vanity, but they delay and interfere with racing, and the car should have been in San Francisco three days ago."[223] But it took four days to drive over good roads from Los Angeles to San Francisco, partly because of the car's poor condition, Scarfoglio wrote. "The springs, the [drive] chains, the engine, the gear, the coach-work, nothing is right, and the engine only works by a miracle which is renewed every day by Haaga, who works through the night, making such repairs as are in his power."[224]

The Zust's elapsed time was 52 days, 2 hours, 20 minutes. The *New York Times* says the Zust and De Dion each traveled 4,090 miles between coasts. As the only car to travel as far south as Los Angeles, however, the Zust traveled farther than the De Dion. Scarfoglio's figure of 4,600 miles is perhaps more reliable.[225] Given its elapsed time, the Zust averaged 3.68 mph and 88 miles per day on its transcontinental journey.

Damage Delays De Dion

Stalled in Cedar Rapids since March 2, the repaired De Dion headed west on March 8. But the French car suffered a "broken driving shaft" when it hit what St. Chaffray called "frozen mud hard as stone" late on March 8 near Le Grand, Iowa "It was found that the repair made in Kendallville was no good," explained St. Chaffray, referring to the broken transmission shaft that stranded the De Dion for three days in Indiana.[226] Disabled again, the car remained overnight in a barn and was towed the few miles to Marshalltown the next day. St. Chaffray boarded a train for Omaha but discovered that the new transmission shaft he had previously ordered from France was held up at customs in New York. He thus had the broken shaft repaired. After a two-day delay, the car left Marshalltown on March 11 to resume the race.

The stretch between Logan and Crescent, Iowa, situated across the Missouri River from and almost in sight of Omaha, was "the hell of Dante" that nearly beat the crew, St. Chaffray wrote. This was the "sinister" Pigeon Creek Valley, "an ascent, narrow in every part, round like an egg, slippery as a soaped pig." Emanuel Lescares "was trying on foot to find in the ocean of dirty mud a place where the car would not completely sink," wrote St. Chaffray, who misspelled the name of the De Dion's newest crew member. "When Lascares came back to his seat we found he had no more rubber shoes. The sticky ground had kept them without warning when he tried to rise to the car.... Lascares was so cold that he did not realize his amusing position."[227]

But Iowans greeted the Frenchmen enthusiastically. "In the streets of towns, men mounted on chairs and holding American flags present to us a temporary triumphal arch. The car must pass between their chairs and under the flags," St. Chaffray said. Another tribute, of sorts, occurred in central Iowa, where "ladies and girls cut with scissors, behind us, the leather straps holding the top and religiously put them on their breasts as souvenirs," St. Chaffray related.[228]

Running late to catch the leaders, the De Dion got stuck in the mud two miles north of Crescent early Friday morning, March 13. At 3:30 a.m. with teams of horses pulling and its engine racing, the car snapped a transmission gear, according to the *Omaha Daily News* and *Omaha Daily Bee*. The disabled car was towed to a Crescent blacksmith's shop; two Omaha garages sent out autos to bring the crewmen and their broken transmission to Omaha.

What Broke?

In contrast to the *Daily News* and *Daily Bee*, the *New York Times* blamed the breakdown on a broken gear in the rear axle. What actually broke? St. Chaffray's *New York Times* dispatches and interviews with Omaha newspapers only compounded the confusion surrounding this question. The damage was to "the pinion in the bevel gear on the shaft," according to the March 14 *Omaha World-Herald*, a vague reference that suggests the rear axle but might also mean the transmission. The same day's *Omaha Daily News*, however, quoted St. Chaffray as saying it was a transmission piece that broke near Crescent. It was the third time it had broken because the De Dion crew had been repairing the break with steel from the United States. But "we have some [steel] from Paris soon," he assured the newspaper.[229]

In the *Omaha World-Herald* account, St. Chaffray theorized that the part in question—broken for the *first*, not the third time—was the result of sabotage by a would-be crewman: "There was a German machinist we had who wanted to go along with us, and when we would not let him he put in a bad piece of gearing—that which has now broken. It is the first time we ever broke this piece, and I can't understand it any other way." Later, while traveling through eastern Nebraska, St. Chaffray told his *New York Times* readers of finding "some screws and a lot of iron pieces in the gear box of the De Dion" while in Iowa. "Some jolly fellows I suppose had jokes with us during our dinner," he said.[230]

The race "has already cost me nearly $10,000 out of my own pocket, but what is the difference so long as I get to Paris first?" St. Chaffray told the *Omaha Daily News*. "It is the honor I am desirous to obtain, and if there is any fair means by which I can win I will do it." In Omaha, St. Chaffray proved "a real spender, too," paying $1.50 for neckties and $8 for shoes, the newspaper reported.[231] While St. Chaffray shopped, machinists at the Union Pacific shops worked around the clock for 2½ days to make a new gear and other parts for the French car. When the parts were ready Sunday morning, March 15, the De Dion crew rushed back to

Crescent, installed them, then drove the revived De Dion across the Missouri River, through Omaha and out onto the Nebraska prairie to run all night.

Breakdowns and Close Calls

Two days later, on March 17, the De Dion got an early start from Grand Island in central Nebraska. St. Chaffray hoped to drive without stopping to the foothills of the Rocky Mountains. Three miles from town, however, a spring broke, and horses towed the car back to Grand Island. Repairs consumed two days. Snow caught up with the Frenchmen near North Platte, where they saw their first Nebraska cowboys, "on horseback and armed with kodaks," said St. Chaffray, who was puzzled by the large number of paths on "unlimited prairies north, south, and west.... Roads sometimes led to isolated farms.... They know, however, of the French car, and all are kind. I am disgusted to have bought a revolver. I send by express my revolver to San Francisco."[232]

On March 23, Day 41, three days after passing Cheyenne (see Fig. 2.38) St. Chaffray tried to race across a small frozen river 10 miles east of Rock Springs, Wyoming, but the heavy De Dion crashed through the ice. Unable to find horses, the crew unloaded the car and wrenched it out with a block and tackle. In the process, St. Chaffray hurt his right hand seriously enough to seek a doctor's care, and speculated about turning the driving chores over to Autran permanently. His hand apparently improved: his *New York Times* dispatches stopped referring to the injury but also revealed that Autran was driving more often.

Fig. 2.38 The De Dion's "towering boxlike body" was gone by the time it reached Cheyenne, discarded in a weight-saving frenzy. Autran is driving, flanked by St. Chaffray. Lescares (wearing a mustache) occupies the back seat. The other back-seat passenger is unidentified. (WDCR)

The next day between Bryan and Granger, Wyoming, the De Dion had two close calls in crossing a stream on a Union Pacific Railroad bridge. A train bore down upon them immediately after the car climbed a steep bank to reach the approach to the bridge. Autran, who was driving, sent the car plunging back to the bottom. The De Dion again climbed the bank and crossed the bridge with the car's left wheels between the rails and right wheels on the ballasted embankment.

"We were still running on the embankment with the deep water on each side and the trestle behind, when we saw far away the smoke of another train approaching us," St. Chaffray recounted. Someone ran ahead to signal the train, which "came to a standstill just one yard from the bonnet of the De Dion."[233] The train then backed up until the De Dion could find a good place to clear the tracks. The De Dion crew slept that night in an abandoned house in the desert.

Dig, Dodge Danger

Near Evanston on March 26, the car got stuck.[234] "At 4 o'clock in the afternoon the De Dion had sunk so badly that the right wheels gave way, and at last the automobile overturned in deep, very deep mud," St. Chaffray recalled. The three men worked in vain with a block and tackle for more than an hour. Salvation appeared in the form of two Hungarian tramps, but they refused to help for money and then "tried to rob us of an axle." Lescares, however, "took out his revolver and gently told the robbers to help us for the sake of their lives. The Hungarians proved to be clever workmen," St. Chaffray wrote.[235] In the light and warmth of cedar fires, the five men tunneled under the car. The mud began to freeze and stiffen but they nevertheless positioned jacks and righted the De Dion. The delay cost them eight hours; they drove to Ogden at night in a snowstorm.

Autran was piloting the De Dion alongside a steep hill east of Cobre, Nevada, on Day 47, March 29, when Lescares in the back seat shouted a warning. Autran braked hard as "a big stone came from the top of the hill and rolled over on us," St. Chaffray wrote in a dispatch from Cobre. "The lower bonnet was half broken. Fortunately the radiator, the motor and the men escaped safely. The broken bonnet was thrown away, and … the motor sent to the open air the song of its four cylinders."[236] But a photograph reveals that the De Dion arrived in Ely, Nevada, later that day with its hood intact—or at least the right side of it. (See Fig. 2.39.) St. Chaffray perhaps meant to say that the rock damaged and the crew discarded the lower left half of the hood.

Their local guide of March 30 appeared so fearsome that Autran whispered to St. Chaffray: "We shall be shot by our pilot," who, it turned out, was preoccupied with other targets. According to St. Chaffray's description,

> This pilot of ours had a revolver with heavy cartridges. He was ugly and dangerous. Each second he was trying his revolver on an antelope, jack rabbits, or cows. He was red with excitement and the pleasure of a trip in an automobile…. He conducted the car by the shortest road without regard to the roughness of said road. Twelve hundred cuts and about six thousand ditches were crossed in the day…. Autran was watching each cut and cried twelve thousand times: "We shall break down."[237]

Fig. 2.39 Occupied only by an unidentified guide, the De Dion draws a crowd in Ely, Nevada, on March 29, 1908. The first man visible to the left behind the guide appears to be Gael S. Hoag of Ely, an ardent Lincoln Highway promoter who traced his interest in good roads to the 1908 New York–Paris race. (NHS)

Lost in Death Valley

The crew crossed the Nevada-California border into Death Valley during the afternoon of April 1. In a sandstorm that raged for eight hours until sundown, Autran lost the trail and the car bogged down. The dazzling sun temporarily blinded Lescares, forcing him to rest in the car. Autran and St. Chaffray shoveled sand and even tore up their shirts to feed under the drive wheels for traction. When the storm abated, they set off on foot and found a prospector camping alongside the trail from which they had strayed.

They bought his tent and commissioned the man to ride his burro to the closest mining camp, Skidoo, 20 miles, where the would-be rescuer was unable to hire teams for the stranded De Dion. But he telephoned Rhyolite, Nevada, which the De Dion had passed on its way to Death Valley. A Rhyolite man who ventured out that evening to travel 50 miles to the stranded car fell from his wagon, broke his neck and died, St. Chaffray recounted.

When his horses returned to camp the next morning, pulling an empty wagon, another wagon started out for the De Dion with food and water. Not necessarily counting on outside help, Autran and St. Chaffray had walked back to their car the previous evening. There, they tore the canvas tent into strips for the car's tires to follow and were nearly out of Death Valley when the Rhyolite rescue team caught up with them, according to St. Chaffray.[238]

The De Dion headed north from Bakersfield on April 4 but traveled just 65 miles before an unspecified accident delayed the crew for 36 hours "on the desert near Tulare."[239] The French car arrived in San Francisco at 5:30 p.m. April 7. "The welcome in the city to the foreign car was so hospitable that we were literally thrown into the arms of a dozen young men, who pressed up close to the footboard of the car," St. Chaffray wrote. "As a sign of joy, and in order to preserve a souvenir of the event, they cut pieces off the canvas hood of the De Dion and took them away as relics."[240] Given its elapsed time between coasts of 55 days, 9 hours, 15 minutes, the De Dion covered 4,090 miles by averaging 3.08 mph and 74 miles per day. (See Table 2.6.)

Table 2.6 U.S. Performance of Autos in 1908 New York–Paris Race

Auto (place)	Elapsed time	Avg. daily mileage	Avg. speed
Thomas (1)	41d-08h-25m	93	3.87 mph
Zust (2)*	52d-02h-20m	88	3.68 mph
De Dion (3)*	55d-09h-15m	74	3.08 mph
Motobloc (disqualified)	–	–	–
Protos (shipped by rail)	–	–	–
Sizaire-Naudin (quit race)	–	–	–

* Their times are similar but the Zust traveled farther than the De Dion. The Zust's average speed and daily mileage are considerably higher in comparison.

Georges Prade, writing in the French *Les Sports*, lamented the poor showing of the De Dion, Motobloc and Sizaire-Naudin:

> It must be confessed, without any pride, that this extraordinary farce is not calculated to raise the prestige of the French automobile industry in the United States, and to help us do business there....
>
> It was not by its speed, whatever may have been said of it, that the American car won; but because its adversaries, ignorant, badly equipped, and awkward, were continually having break-downs....
>
> We think that French industry merited better; but wishes are not realities, and the very simple and very sad reality is that the three French makes of cars engaged in this performance have spent a great deal of money to cause a laugh, both at us and them. This is what may be called furnishing a boot wherewith to kick us.[241]

Gratefully receiving the French boot, *Automobile Topics* eagerly used it: "If it were not for the American entry in the New York–Paris race there wouldn't be very much left of that much advertised event. For a rank outsider, a car that French motorists looked upon with mingled indifference and contempt, to step in and capture all the glory is a trifle exasperating when you come to think of it."[242]

Stragglers Fight Mud, Breakdowns

The Protos, eight days behind the Thomas in leaving Chicago, spent all or part of nine days in crossing Iowa, compared to five days for the Americans. Part of this time was spent idle in Ames, where the Germans ran out of tires. Because of tire and mechanical problems, the Protos took nearly two weeks apiece in crossing Wyoming and Utah. (See Table 2.7.) The Motobloc left Chicago on March 7, three hours ahead of the Protos, but quickly fell behind the German auto. Desperate to catch the leaders, Godard eventually shipped his disabled Motobloc by rail to the West Coast, which disqualified him from the race.

Table 2.7 Full or Partial Days Consumed Crossing Western States

	Iowa	Neb.	Wyo.	Utah	Nev.	Calif.[1]
De Dion	15	5/6[2]	7	3	5	7
Protos	9	5	13	14	–	–
Thomas	5	5	7	4	5	4
Zust	7	6	10	5	5	7

[1] To reach San Francisco; this doesn't include the car's stay there.
[2] The De Dion entered Nebraska either late on March 15 or early in the morning on March 16, 1908.

Sources: Based on an itinerary compiled from articles and racers' dispatches in the *New York Times*; local newspapers; Antonio Scarfoglio (Zust), *Round the World in a Motor-Car*, and George Schuster (Thomas), *The Longest Auto Race*.

The first day out of Chicago, while running through what Godard estimated was three feet of mud near Maple Park, the Motobloc's drive chain picked up a stone and fed it into the sprocket wheel, ruining a ball bearing. The next day, March 8, the Protos towed the French car nine miles to DeKalb, where Godard, in defiance of race officials, loaded the Motobloc on a train and shipped it 80 miles to Clinton, Iowa, for repairs that were completed March 12. Godard also replaced his hard-rubber "cushion" tires with pneumatics. Meanwhile, the Protos went on ahead in fourth place.

To avoid delays in shipping the car back to DeKalb by rail, Godard made up his 80-mile advantage by driving the Motobloc the 40 miles to Dixon, Illinois, and then back to Clinton. He reached Cedar Rapids, Iowa, on March 13, but got lost trying to find Marshalltown the following day. A Cedar Rapids motorist saved the day by redirecting Godard, dripping wet from a rainstorm and "wandering miles off the route."[243] He reached Tama—50 miles west of Cedar Rapids and short of his Marshalltown destination—at midnight on March 14 after 15 hours of hard driving. (See Fig. 2.40.)

In the Iowa mud on March 15, "my car sank in the road almost out of sight," said Godard, who spent the night at a farmhouse two miles east of Ames. "We are hopeful we may some day come out of it all and run once more on solid ground."[244] That hope dissolved on the following day when hub-deep mud held Godard's progress to the 25 miles from near Ames to Ogden, Iowa. The last eight miles took three hours.

Fig. 2.40 The Motobloc upon finally reaching Marshalltown, Iowa. Left to right, three men in front: unidentified, Godard and Hue. Standing in back: unidentified and Livier. (LOC LC-USZ62-37967)

"In The Baggage Coach Ahead"

Between Chicago and San Francisco, a dispute arose among the racers. At issue: did the race rules allow contestants to ship their cars to the West Coast so as to reach Alaska before the spring thaw? When they ran into bad eastern roads, the crews struck a gentlemen's agreement "whereby it was understood that every car should do its level best until March 5 or 6—just as it suited—and then cars should be sent by train to Seattle" and shipped to Alaska, the *Chicago*

Daily Tribune reported. "On the presumption that this agreement was still in effect the Germans in the Protos and the French in the Moto Bloc didn't hurry much in reaching Chicago, figuring that they could ship their cars from this city."[245] But in answer to Koeppen of the Protos, officials with both *Le Matin* of Paris and the *New York Times* sent telegrams directing the racers to continue west under their own power.

After an unspecified breakdown outside of Carroll, Iowa, March 17, Godard—out of money, parts and patience—shipped the Motobloc to Council Bluffs and booked it through to San Francisco on a Union Pacific flatcar. *Le Matin* officials wired him their approval, Godard insisted to an Omaha newspaper. "The rules of the contest which allow this act are not made public," sneered the *Omaha World-Herald*. The newspaper suggested that a band should greet the Motobloc at the Council Bluffs train station by playing "In the Baggage Coach Ahead."[246]

"We have been on the way five weeks and have spent or strewn dollars by the road because the car has at each moment needed assistance to get out of the mire, snow, and mud in innumerable instances," Godard replied. "I cannot continue to carry along the Motobloc on account of charges unforeseen and without end." Interpreter L.J. Cheramy of Clinton, Iowa, said the French-speaking crewmen did not have an interpreter with them in New York City when they had extra parts shipped ahead. The parts were mistakenly directed to the West Coast instead of cached along their route, "thus shattering all hopes of the crew fixing their machinery and proceeding with the race by its own power," as the *Omaha World-Herald* put it.[247]

Tires Trouble Teutons

Koeppen's daily strategy for the Chicago-San Francisco leg was to leave early, skip lunch and travel till dark. The Protos reached Cedar Rapids, Iowa, early on March 11, five days out of Chicago. "The car has exhausted its extra supply of tires which it carried at the start, and its surplus supply has been sent to Seattle," according to a news report from Cedar Rapids.[248] A photo of the Protos in a Cedar Rapids garage shows one spare tire—useable or not?—hanging from the car, along with two empty rims. (See Fig. 2.41.) The two-man crew spent a day in Cedar Rapids to repair tires, ship ahead some extra weight and brace the springs in an effort to increase road clearance and save the tires.

On March 13, the "materially lightened" Protos traveled just 38 miles between Marshalltown and Ames, where it was stranded for $2\frac{1}{2}$ days. (See Fig. 2.42.) "A tire exploded at Nevada," outside of Ames, and the car sprung an axle between Nevada and Ames, the *Omaha Daily Bee* reported.[249] The frustration of waiting in Ames provided Koeppen with dozens of moments "in which one feels like shooting out of one's skin with a loud report, like the bursting of a thousand pneumatic tires," he wrote in his book.[250] Koeppen had new tires shipped from Chicago and ordered tires from his Seattle cache shipped to Grand Island.

The German car that rolled into Omaha on March 17 looked to be the acme of preparation, as the machine "is entirely plastered with extra parts in case of breakdowns," noted the *Omaha*

Fig. 2.41 The Protos in Cedar Rapids, Iowa. (Im Auto um die Welt)

World-Herald. But like other foreign racers, Koeppen admitted, he suffered from a "lack of appreciation of the difficulties of the road." For example, he had come nowhere close to his expected average of 100 to 150 miles daily. "Our main trouble has been with tires. We should have had relays of tires on the road for us at various places, but we did not know what we faced. The supply which [w]e counted on taking us across the country … took us only to Ames."[251]

Before leaving Omaha, mechanics at the Kimball garage replaced the car's second-speed gear. Two days later, on March 19, the Protos picked up its new tires in Grand Island, where it arrived four hours after the De Dion departed. Hundreds of cheering spectators surrounded the car when it reached Cheyenne on March 21, trailing the Thomas by 14 days, the Zust by 10 days and the De Dion by a day. "The machine bears on its sides thousands of signatures of spectators who saw it pass through the various cities visited in its 2,000-mile run from New York, and Cheyenne furnished its quota to the strange collection of autographs," reported a wire story from Cheyenne. Commenting on the same phenomenon, Schuster of the Thomas recalled that "at nearly every stop somebody from the crowd would write or scratch his name on the car."[252]

Thayer Today, Ogden Tomorrow

Driving snow, high winds and muddy, slippery roads hindered Snyder, the Protos driver, between Cheyenne and Rawlins during the following three days. The Protos stalled while

126

Fig. 2.42 Marshalltown's brick streets give the Protos a respite from Iowa mud. Snyder is driving, Koeppen—shorn of his moustache—beside him. The third man is unidentified. Large chunks of tread have torn away from the studded rear tire; one headlamp has lost its glass. (LOC LC-USZ62-37968)

fording Rock Creek on March 24. And then, the car broke down. According to conflicting reports, the car on March 26 broke either a "driving shaft" or "crank shaft" in the mud at Thayer Junction, a few miles west of Point of Rocks, Wyoming. Most likely it was a driving shaft—possibly a transmission part—because a broken engine crankshaft would have likely forced the auto from the race. The German crew shipped the Protos some 200 miles ahead by rail for repairs at the Union Pacific shops in Ogden, Utah, and returned to Thayer Junction on March 30.

But new snow and the previous week's thaw had left the roads a quagmire. It took Snyder four days to drive the Protos back to Ogden. He arrived April 2, calling the roads between Granger and Ogden nearly impassable, and complaining about the unexpectedly high cost of crossing the United States. "Mein Gott," he exclaimed to an *Ogden Standard* reporter, "it is no race, but an expense endurance test."[253] A Utah State Historical Society photo depicts the Protos as it left the Ogden Automobile Company at 10:30 a.m. April 3. The photo caption says Koeppen earlier that morning took the train for Lucin, Utah, leaving the car in charge of Snyder and guide Charles Orson Wheat of Ogden. (See Fig. 2.43.)

Fig. 2.43 Snyder, left, and local guide Charles Orson Wheat prepare to leave Ogden for Brigham City, Utah, on April 3, 1908. (USHS)

Perhaps the strain of driving through mud caused the broken differential that halted the car in Kelton, Utah, on April 4 and 5. An unspecified engine problem stranded the car 20 miles west of Kelton on April 7. The Union Pacific complied with Koeppen's request to send a special train over "the old disused Central Pacific tracks to carry the machine back to Ogden," which it did on April 8.[254]

Koeppen left immediately for Seattle by train and returned to Ogden with repair parts—a new cylinder, according to *Motor Age*[255]—on April 10. But news from Valdez sent him hurrying back to Seattle, according to some news accounts. According to his book, however, Koeppen left Ogden by train on the night of April 7, arrived in Seattle at 6 p.m. on April 9 and didn't return to Ogden.

Despite the *New York Times'* earlier scouting, the Americans on April 9 tested the snow-packed trail at Valdez and found it too narrow and too soft for a car. The same day they loaded the Thomas aboard a Seattle-bound steamer; organizers subsequently ruled that the race to Paris would continue from Vladivostok, Siberia, in the Russian Empire.

Onward to Paris

Machinists at the Ogden UP shops finished repairing the Protos on April 11 while Koeppen waited in Seattle for instructions from abroad on how to proceed, according to the *New York Times*. In his book, Koeppen said he also met with St. Chaffray, who had arrived in Seattle ahead of his De Dion, to discuss the change in route. On April 14, the De Dion and Zust left Seattle by steamship, getting a jump on the Thomas in reaching Siberia to resume the overland race. Later in the day, Koeppen announced that he was forced to ship the Protos from Ogden to Seattle by rail. (See Table 2.8.)

In fact, the Protos left Ogden by rail on April 12, strapped to a flatcar, and passed through Pocatello, Idaho, the following day, as an unimpressed *Pocatello Tribune* reported: "It is a big machine, but in its present condition looks like a pile of scrap iron."[256]

Koeppen, too, left Seattle ahead of the Thomas. Shipping by rail would disqualify him, as it had disqualified Godard's Motobloc, but Koeppen said he would continue the race as a noncontestant. As it turned out, however, St. Chaffray, commissionaire general of the race, had given Koeppen permission to ship the car. *Le Matin* thus allowed the Protos to remain in the race.[257] But race organizers granted the Thomas an allowance of 15 days to compensate it for its lead into San Francisco and its detour to Alaska. Officials also penalized the Protos crew 15 days for shipping its car from Ogden to Seattle.[258] This, in effect, gave the Thomas a 30-day lead over the Protos and a 15-day lead over the other contestants.

What happened to the other racers? The Sizaire-Naudin, of course, quit early, in New York state; Godard dropped out at San Francisco—in response to a factory telegram to "Abandon race, sell car and come home"—where he sold the Motobloc for $1,650[259]; and the Werner, the *New York World's* rebel racer, quit at Seattle in late April, having traveled much of the way from New York by railcar.

The De Dion likewise dropped out, at Vladivostok, the starting point for the race across the Asian mainland—leaving only the Protos, Thomas and Zust. With a new crew, Koeppen and the Protos traversed the estimated 8,000- to 10,000-mile route across Asia and Europe to reach Paris first on July 26. The Thomas arrived four days later and the Zust limped in on September 17.[260] Because the time allowance and penalty against the Protos had given the Americans a 30-day lead, the Thomas was declared the winner of the 1908 New York–Paris race.

Thomas Flyer "Still Going"

Later, the automaker used its round-the-world Thomas Flyer mainly for advertising. Thus it appeared at various auto shows. The *Scientific American* of November 14, 1908, pictures the car pulling a glider aloft, and in 1909 the auto traveled nearly across the country again in plotting

Table 2.8 Daily Progress and Position of the Racers in the United States from March 17 to April 12, 1908*

Day/Date	De Dion	Motobloc	Protos	Thomas	Zust
35 - 3/17	Grand Island, Neb.-3	Carroll, Iowa-5	Omaha, Neb.-4	Cobre, Nev.-1	Granger, Wyo.-2
36 - 3/18	Grand Island, Neb.-3	Omaha, Neb. [1]-5	Columbus, Neb.-4	Veteran M.C., Nev. [2]-1	W Granger, Wyo.-2
37 - 3/19	W Ogallala, Neb.-3	disqualified [3]	Lexington, Neb.-4	Twin Springs, Nev.-1	Spring Valley, Wyo.-2
38 - 3/20	Cheyenne, Wyo.-3		Ogallala, Neb.-4	Tonopah, Nev.-1	Evanston, Wyo.-2
39 - 3/21	Medicine Bow, Wyo.-3		Cheyenne, Wyo.-4	Death Valley, Calif.-1	Ogden, Utah-2
40 - 3/22	Bitter Creek, Wyo.-3		Laramie, Wyo.-4	E Bakersfield, Calif.-1	Kelton, Utah-2
41 - 3/23	Bryan, Wyo.-3		Rock River, Wyo.-4	ca. Los Banos, Calif.-1	Ogden, Utah[4]-2
42 - 3/24	Church Buttes, Wyo.-3		Rawlins, Wyo.-4	San Francisco-1	Ogden, Utah-2
43 - 3/25	Carter/Spring Valley, Wyo.[5]-3		W Green River, Wyo.-4		ca. Cobre, Nev.-2
44 - 3/26	E Ogden, Utah-3		Thayer Junction, Wyo.-4		Ely, Nev.-2
45 - 3/27	Ogden, Utah-3		Ogden, Utah[6]-4		Twin Springs, Nev.-2
46 - 3/28	Montello, Nev.-3		Ogden, Utah-4		Goldfield, Nev.-2
47 - 3/29	Ely, Nev.-3		Ogden, Utah-4		ca. Death Valley, Calif.-2
48 - 3/30	Tonopah, Nev.-3		Granger, Wyo.-4		Daggett, Calif.-2
49 - 3/31	Goldfield, Nev.-3		W Carter, Wyo.-4		Los Angeles-2
50 - 4/1	W Stovepipe Wells, Calif.-3		Evanston, Wyo.-4		N Santa Barbara, Calif.-2
51 - 4/2	Death Valley, Calif.-3		Ogden, Utah-4		S Bradley, Calif.-2
52 - 4/3	ca. Bakersfield, Calif.-3		Kelton, Utah-4		S San Francisco[7]-2
53 - 4/4	Tulare, Calif.-3		Kelton, Utah-4		San Francisco-2
54 - 4/5	Tulare, Calif.-3		Kelton, Utah-4		
55 - 4/6	N Fresno, Calif.-3		Kelton, Utah-4		
56 - 4/7	San Francisco-3		W Kelton, Utah-4		
57 - 4/8			Ogden, Utah[8]-4		
58 - 4/9			Ogden, Utah-4		
59 - 4/10			Ogden, Utah-4		
60 - 4/11			Ogden, Utah-4		
61 - 4/12			shipped by rail[9]		

* As of midnight each day. Because the racers often traveled all night, this represents each car's last known daily position, as reported in newspapers and in the crewmen's own accounts.

1 Shipped by rail.

2 Veteran Mining Camp was just a few miles southwest of Ely, Nev., according to driver Harold S. Brinker.

3 Disqualified for shipping car from Omaha to San Francisco by rail.

4 Shipped by rail to Ogden for repairs and later returned to Kelton by rail.

5 G. Bourcier St. Chaffray gives conflicting accounts of his night stop.

6 Shipped by rail to Ogden for repairs and later returned by rail to Wyoming.

7 Antonio Scarfoglio says the Zust stopped in Ramsey, Calif., a city that did not appear in a search of several geographical sources and indexes at the California State Library. The *New York Times* names San Jose as the night stop.

8 Shipped by rail to Ogden for repairs.

9 Hans Koeppen shipped the Protos by rail from Ogden to Seattle. Race organizers could have disqualified him, but it turns out Koeppen acted with the approval of *Le Matin* representative G. Bourcier St. Chaffray. The Protos was penalized but allowed to remain in the race.

a route for the New York-Seattle race. When it arrived in Seattle on May 19, 1909, its odometer registered 36,979 miles, according to the *Seattle Post-Intelligencer*. "New York-to-Paris Winner Still Going," proclaimed a headline in the April 25, 1912, *Automobile*. The article said two factory "service men" were driving the car from Buffalo into the New England states "to assist the Thomas dealers in the inspection of owners' cars."[261]

Industrialist and newspaper publisher C.A. Finnegan acquired the car during a March 1913 auction of the Thomas factory and its contents. The Thomas Flyer "rusted for years in the garage of his home at Elma, New York," just outside Buffalo, according to Schuster.[262] Finnegan sold the Flyer in 1947 and in 1948 the new owner resold it to automotive historian and collector Henry Austin Clark Jr. It remained in Clark's Long Island (New York) Auto-motive Museum until 1963, when Harrah's Automobile Collection of Reno, Nevada, bought it and restored it to appear as it did upon arriving in Paris. The William F. Harrah Foundation's National Automobile Museum in Reno now displays the car and the trophy it won.[263] At least one other New York–Paris racer survives. The restored Protos—minus the special body that had been fitted behind the front seat for the 1908 race—is preserved at the Deutsches Museum in Munich, Germany.[264]

Men and Makers Pass On

Of the Thomas drivers, Brinker died in 1955 at age 69, Roberts in 1957 at age 74, and Schuster, proclaimed "the last survivor of the longest, maddest automobile race in history,"[265] in 1972 at age 99. He thus lived to see and even drive the restored Thomas Flyer. Reporter T. Walter Williams, who traveled with the car as far west as Iowa, died in 1942 at age 77. Mathewson's date of death could not be determined.

Autos bearing the names of De Dion, Motobloc, Protos, Sizaire-Naudin and Zust all dis-appeared from the scene before or during the 1930s. Ironically, the maker of the self-styled "Champion Endurance Car of the World," which spent an estimated $100,000 in the New York–Paris race[266], was the first to fail. "Bad judgment and bad luck" kept the Thomas company from growing much after the 1908 race, according to Schuster, who worked for the E.R. Thomas Motor Company until its demise. Bad cars had something to do with it, as well, because the 1909 models were "noisy, underpowered, and literally leaked oil," he said.[267] By refusing to acknowledge that the Thomas Flyer suffered many of the same breakdowns as the foreign autos in the Great Race, the manufacturer has-tened its own demise, Schuster contended:

> One factor, I have always believed, was an utterly misleading booklet that the company published on the around-the-world race. It left out all our troubles with the car, and said that it "was never in a repair shop" and "none of the valves were ground or changed; not a sparkplug was changed; nor were the crankshaft bearings changed or adjusted."

Many people knew differently, and I protested to no avail. Our booklet appeared almost a joke in comparison with the long and candid books by men on the competing German and Italian cars, which appeared at the same time and were translated into many languages.[268]

The Legacy of the Race

The American leg of the New York–Paris trek generated praise and jibes—sometimes from the same source. "Where is the bard who will sing to the world in fitting measures the race of the automobiles from New York to Paris?" asked *Current Literature*, inspired by the audacity of the enterprise. But the monthly magazine was skeptical of the *New York Times'* claim that the automobile race "will show to the world that here at last is a power greater than any of the wonderful means of transportation which human ingenuity has yet given us." Since that proclamation, *Current Literature* noted,

> the cars have been towed by horses through innumerable snow-drifts, carried on stone-boats over other drifts and dug out of countless snow-banks where they would be today if they had had to depend on their own power. The automobile is a wonderful achievement of mechanical skill, but the horse is still an important adjunct of civilization. The impression that the horse is disappearing from view is very erroneous.[269]

Outing Magazine, without mentioning horses, concluded that the performance of the New York–Paris racers "goes to prove that the automobile has passed entirely out of the experimental or play-thing stage, and is a dependable machine for work as well as for pleasure."[270] The race across America was still in progress when San Francisco-based *Overland Monthly* declared that "the wonderful race that the cars are making ... has opened the eyes of the skeptical to the great power and endurance of the modern automobile."[271]

Though its long-term effects are harder to measure, the transcontinental portion of the New York–Paris race inspired in spectators at least a temporary appreciation for, if not awe of, the automobile. The contest "is certainly waking up interest in automobiles over the route we have come," Roberts observed in Iowa. The·*Omaha World-Herald* noted that the race caused a local sales boom: "Dealers say that their sales have increased mightily and that they are constantly besieged with prospective buyers asking for demonstrations."[272]

The racers also provided inspiration for daily living, as one preacher sermonized: "In this great automobile race the reward will come to the men who patiently persevere in the face of gigantic obstacles. This quality is essential also in the running of the race of life."[273]

Generations later, the round-the-world race still symbolized a human and mechanical triumph in the face of adversity. In 1952, a Spicer Manufacturing ad pictured the Thomas Flyer and proclaimed it "the world's first round-the-world car[,] equipped with a Brown-Lipe transmission built in the Spicer pattern of precision!" In a 1994 *Fortune* ad, the Siemens

Corporation asserted that it built the Protos, "one of the favorites in The Great New York to Paris Race," and that in 1994 it was building "almost everything for an automobile except the automobile itself."[274]

Years later, Williams commented on the implications of the 1908 race:

> The single fact that three automobiles [De Dion, Thomas and Zust] were able to run from New York to San Francisco in winter under their own power and at high speed, everything considered, shed new light upon the sturdiness and power of the motor car and gave it a tremendous lot of publicity in the public prints.[275] [See Fig. 2.44.]

And what happened because of the "daily, wide, and effective advertising" the race had given to the primitive condition of America's roads? Very little, according to those automobile

Fig. 2.44 These and other headlines in the public prints traced the progress of the New York–Paris racers through the United States.

owners who wanted to see immediate changes. But the race did influence a great many policy makers and others—among them Gael S. Hoag of Ely, Nevada, one of the builders of the first U.S. transcontinental auto route, the Lincoln Highway. Hoag "attributed to this race the awakening of his interest in highways," according to the Lincoln Highway Association.[276]

Likewise, Arthur R. Pardington, the future vice president and secretary of the Lincoln Highway Association, avidly followed the New York–Paris race. As Pardington predicted before the start:

> The cars that finish—and I would also include all those that get through Alaska and enter Siberia, whether they break down after that or not—will surely furnish the most remarkable object lesson of the adaptability of the modern motor car and the intelligent driver to accomplish literally anything that they set out to do.[277]

Distant countries seemed a little less distant after the round-the-world race, if only because the race had served as a practical geography lesson for perhaps millions of American schoolchildren. According to the superintendent of 60 schools where pupils studied the race daily,

> One child will be asked to tell what the Thomas car did the day before, where it started from, what point it reached, what cities it passed through, and the rivers it crossed and how. If this younger narrator goes wrong by a hair's breadth up go several hands. The children are permitted to thus correct each other, thereby learning to be attentive and accurate, and unconsciously they are imbibing geography and the language at the same time.[278]

Another Transcontinental Race

In May, while the New York–Paris contest was still on, the *New York Times* proposed a race from New York City to San Francisco and back—open only to stock cars. Wiser now, the newspaper published a list of 13 rules that set weight and horsepower standards, prohibited crew changes and placed in each auto an official observer, who would issue penalties for every repair made en route. According to the *Times*, these rules would ensure that

> [t]he winner will be, not the car that brings back to the starting point little more than its name, owing to repairing and substitution of parts on the way, but the car that can most quickly twice cross the continent because of the excellence of its original construction.[279]

A score of automakers who met at the *New York Times* office to offer suggestions gave the proposed race their "heartiest indorsement," as the newspaper put it. Charles Glidden, who lent his name to the Glidden Tour, a popular annual endurance contest, hailed the new race as "one of the most attractive events and tests of motor car durability." Montague Roberts would drive one of the two dozen or more cars that were expected to enter. "It ought to arouse a bigger public interest than the New York to Paris race, which was certainly the most popular competition ever held in automobiling," Roberts predicted.[280] By August, nine automobiles had officially entered the race, which would start from Times Square on

July 4, 1909: Gearless, Lozier, Thomas and two Maxwells, all from America; and Fiat, Isotta, Renault and Zust, all from Italy.

The race never occurred. When the Manufacturers' Contest Association met on March 30, 1909, and refused to sanction the contest, the *New York Times* canceled the event.[281] But another unsanctioned contest—a race from New York to Seattle—filled the void that summer. And while the New York-Seattle race fell short of the New York–Paris trek as an act of splendid folly, it was fraught with adventures, difficulties and lessons all its own.

CHAPTER 3

New York to Seattle: A Wild Race across the Continent

MARSHALL, Mo., June 14—The Itala car, fifth in the ocean-to-ocean automobile race, arrived today, considerably damaged. While crossing the Missouri River on the railroad bridge at Glasgow the car was struck by a freight train, which knocked off the gasoline tank and broke the front axle.

A year after the 1908 New York–Paris contest, mining magnate M. Robert Guggenheim offered a bag of gold to the winner of a race from New York City to the Alaska-Yukon-Pacific Exposition in Seattle. Guggenheim, a good-roads enthusiast and colorful showman, engaged the winner of the 1908 race to blaze the trail that the 1909 racers would follow. Hoping for 20 starters, he offered a $2,000 trophy, $5,000 in gold and an incalculable amount of free publicity to the winners of the race.

Guggenheim was after a prize of his own—turning the race route into the nation's first transcontinental highway. The prospects were promising: Several states appropriated money to improve the route and President William Howard Taft himself lent his support by agreeing to start the race, which attracted 13 official entries. As the June 1 starting date approached, however, many entrants withdrew from the contest, as they had done before the daunting 1908 around-the-world race. This time, however, a powerful group of "licensed" automakers thinned the ranks by opposing the New York–Seattle contest, fearing it would become an unfettered "wild race across the continent."

Consequently, "no race ever run encountered one-tenth as much opposition as did this," claimed the Ford Motor Company.[1] Henry Ford gleefully entered two Model T's, anxious to snub the group representing licensed manufacturers —those who paid royalties under the Selden patent. George B. Selden in 1895 received the first patent covering a complete automobile, and the patent holders asserted that every auto made since was infringing on that patent. Disputing this claim, Ford and others refused to pay royalties. In a 1911 courtroom victory, Ford broke what he called the patent's stranglehold on the auto industry.

In 1909, however, he sought a victory over the licensed manufacturers—that is, members of the Association of Licensed Automobile Manufacturers (ALAM)—in the grandly titled International Contest for the M. Robert Guggenheim Transcontinental Trophy

137

and Cash Prizes. The real reason none of the licensed makers entered, Ford claimed in *The Story of the Race*, his company's post-race publicity booklet, was "a knowledge of the inability of the product to go through creditably."[2]

The maneuvering during the race was just as exciting. Even while laying out the route, the Thomas Flyer, winner of the 1908 New York–Paris race, suffered a severe crack-up, signaling the rigors that awaited the official entrants. Opposing the two lightweight Fords were three heavier and more powerful cars—an Acme and a Shawmut, both American autos, and an Itala, the only foreign entry. In Chicago, the Itala's exhausted driver checked into a hospital. His hapless teammates, racing to catch the leaders, first overturned the car in an Illinois ditch and later tangled with a freight train in Missouri, where shotgun-toting farmers warned the Fords to slow down—or else.

The racers all plowed through deep mud from Illinois to Wyoming. Lost in the mire, the Acme crew drove around aimlessly for part of a day before finding its way into Cheyenne. Similarly disoriented, Ford No. 1 lost the lead and nearly two days in the Northwest. Comfortably ahead and within 300 miles of Seattle, Ford No. 2 caught fire at a refueling stop, which destroyed the gasoline tank and charred parts of the car. The drivers pressed on, feeding the engine from a gasoline can at their feet. Though lacking money, spare parts and a network of local dealers, the Shawmut performed heroically, according to Henry Ford, who vowed to buy one.

A crowd of 10,000 people watched the finish of the fastest transcontinental auto race of the era. Yet the contest was far from over, for the owner of the second-place auto accused the first- and third-place finishers of cheating. The fourth-place team leveled the same charge against the first three finishers. Among the accusations: Racers had illegally replaced parts, taken aboard unauthorized drivers and even bribed ferryboat operators to delay the other cars. It wasn't just sour grapes—some of the charges were true, prompting Guggenheim, as referee, to disqualify one laggard finisher. Then, despite a formal protest, he awarded the trophy and cash prize to the first-place finisher. A committee later nullified the victory, disqualified the apparent winner and awarded first prize in the Great Race of 1909 to the second-place auto.

"A Big Motoring Contest"

According to J.E. Chilberg, president of the Alaska-Yukon-Pacific Exposition,

> The object of the ocean to ocean contest is to call the attention of the people of the United States to the latest achievements of the Northwest, to encourage the making of good roads in that territory and to mark the progress which has been made since the first caravan creeped over the plains in search of the wealth of the Pacific coast.[3]

Less charitably, *Motor World* called the race "a method of 'working' the newspapers for free publicity calculated to assist the Seattle exposition."[4] Guggenheim, 24, was involved in the exposition at least partly because his family's mining companies were unearthing much of the wealth Chilberg mentioned. Guggenheim would eventually become a director of the American Smelting and Refining Company and an officer of the U.S. Zinc Company, as well as a U.S. ambassador to Portugal.[5] The projected highway, in addition, would link Guggenheim's "old home, New York, and his new home, Seattle."[6]

Expo officials and the Seattle Automobile Club were involved in the planning, which began late in 1908, according to a February 18, 1909, announcement in *Motor Age*:

> The American continent will be the scene of a big motoring contest during the spring and early summer.... Already letters have been sent out to all motor car manufacturers giving them an opportunity to lay out the route....

> The prize for the contest will be a handsome trophy which ... will be given outright to the winner and from present plans the winner will be the car that reaches Seattle first, having traveled over the road on its own wheels without any other rules or conditions being imposed....

> Tom Moore, of Mills & Moore, who will be in charge of the details of the race, has already left [New York City] for Seattle ... in order to arrange the rules and entry blank.[7]

Motor World praised the "beautiful simplicity" of the rules. In March, race organizers announced that they would send an agent over the route "to arrange for the suspension of speed laws in the counties and cities through which the route will be made."[8] The agent apparently failed. Citing Eastern speed limits, race organizers settled for making the transcontinental spectacle an endurance contest to St. Louis, and from there an over-the-road race to the West Coast.

Automakers Oppose Race

Seeking a trophy worthy of the car and crew tough enough to win a transcontinental race, Guggenheim offered $250 "for the best design to be submitted by artists of the northwest." From 116 entries, he selected a design submitted by Shreve & Company of San Francisco, for a 42-inch-tall, $2,000 trophy.[9] (See Fig. 3.1.) "The gold cover will represent the Northern Hemisphere, on which will be poised a flying wheel supporting the figure of Victory, draped with a banner bearing the inscription 'New York to Seattle.'"[10] Guggenheim also offered cash prizes of $2,000 for first place, $1,500 for second, $1,000 for third, $500 for fourth and $300 for fifth. Since the finish line was on the grounds of Seattle's Yukon-Pacific Exposition, the prize money "will be paid to the winners in nuggets of Yukon gold."[11]

Once Guggenheim put up enough gold to guarantee the cash prizes, the Automobile Club of America—which had jurisdiction over American motor contests having an "international character"—sanctioned the event. But the Manufacturers' Contest Association (MCA) did not. In February 1909, automakers and importers with racing programs formed the MCA to

Fig. 3.1 The Guggenheim trophy. (UWSC Nowell X673)

help shape rules governing racing. According to MCA President Benjamin Briscoe of the Maxwell-Briscoe Motor Company, the new group would advise the American Automobile Association, which sanctioned and policed U.S. racing.[12] MCA members included some of the biggest foreign and domestic automakers—Buick, Chalmers-Detroit, Hudson, Isotta, Knox, Lancia, Locomobile, Lozier, Marmon, Maxwell, Packard, Peerless, Premier, Renault, Reo, Stearns, Stoddard-Dayton, White and Winton.[13]

The group supported contests "that furnish to the public a basis of the comparative merits of cars and that demonstrate to the manufacturers where improvements can be made in material and construction." The MCA, however, announced on March 30, 1909, that it opposed the speeding that the New York–Seattle drivers were liable to engage in. Allowing day-and-night driving would turn the event into a endurance contest not for cars but for drivers, the group argued. What's more, the race rules as printed on early entry blanks, "which among other things provide that engines, axles, gears, &c., may be changed twice, and only the original frame brought to Seattle, are not for the betterment of motor car manufacture."[14]

Race organizers tried to pacify the group. "That the contest will be a wild race across the continent is denied … by its managers, who point out that in the entry blank contestants are cautioned against exceeding the speed limit of any town through which they may pass," the *New York Times* said. Besides, Chilberg argued in a letter to Ohio Governor Judson Harmon, the cars will "carry such a load and be equipped in such a manner that excessive speed will be impossible." And Guggenheim reacted by announcing "that he, as referee, will see that the speed laws of the various States through which the cars will pass are not violated."[15]

Entries Trickle In

Despite the MCA's opposition, a Simplex and privately-owned Stearns entered the race, followed in order by two Fords and an Acme. First, on April 2, the Simplex Automobile Company of New York City announced it would use a 50-horsepower "regular stock model." Specially geared for hill-climbing and weighing 6,000 pounds with equipment, it would have a top speed of just 40 mph. The automaker hoped to hire racer George Robertson as driver. In so reporting, the *New York Times* added that "a privately owned Simplex will engage in the contest also, the owner driving and being accompanied by his friends."[16] By mid-April, the newspaper was referring to a single Simplex entry. In any case, no Simplexes started.

Oscar Stolp of Sheppard & Stolp in New York City, a maker of wire goods, entered his own Stearns and asked for help from the F.B. Stearns Company of Cleveland. The automaker, however, belonged to the Manufacturers' Contest Association, which was boycotting the race. F.B. Stearns personally refused Stolp's request for a mechanic, financial backing and repair stations along the route, and "strongly" advised him to quit the race. "I assure you that for a private owner, the trip is impossible," Stearns wrote in a letter to Stolp. "How can you hope to compete with fully equipped factory organizations who are willing to spend $15,000 or

$20,000 to get their car to Seattle in record time?" In reply, a spokesman said Stolp "regrets that the Stearns people have not sufficient confidence in their car to back up its performance."[17]

F.B. Stearns' cost estimates stirred up a debate within his own company—a reflection of how the race had created divisions throughout the auto industry. "There has been considerable friction over the ocean-to-ocean contest," the *New York Times* delicately put it. Taking issue with his boss was C.F. Wyckoff, the Stearns company's eastern representative. "Mr. Stearns has created the wrong impression," Wyckoff told a reporter. It may cost $20,000 to prepare a special car for track racing. But as for preparing for a transcontinental endurance contest, "there is no reason why the cost should be more than $2,000 at the outside," Wyckoff said. "It should be just the same as touring 4,000 miles in New York State."[18]

In an editorial titled "Shouldn't Be a Race of Dollars," the *New York Times* agreed: "To be really significant, from the purchaser's point of view ... every dollar spent for repairs on the road should count adversely in the score, and to have crossed the continent cheaply, as well as at reasonable speed, would be a vastly better advertisement for a car than a demonstration of amazing swiftness."[19]

Ford: "The Best Car Should Win"

Henry Ford, who on April 11 entered two 20-horsepower Model T's, sided with Guggenheim and other race supporters, in defiance of the MCA:

> This will be the first real auto contest ever promoted. It will give Americans an opportunity to appreciate the vast possibilities of the motor car. The contest will show manufacturers the weak points in auto construction ... and the buyer will learn more about motor cars than he could in any other way.
>
> The rules governing this contest are as fair as could be devised. The best car should win, which is not generally the case in the average race or contest. I'm going in to win, of course, but there is another reason why I have entered two cars. This race will be instructive to me. It will show up the weak points of the cars.... It seems to me that every manufacturer having faith in his product should welcome this chance to prove it.[20]

Likewise, H.M. Sternbergh, president of the Acme Motor Car Company of Reading, Pennsylvania, regarded the race as

> the most important and impressive of any automobile competition ever promoted in this or any other country. Not only will it attract widespread interest to the automobile and demonstrate the durability and strength of American cars, but it will call very general attention to the execrable road conditions throughout the country.
>
> It is a source of everlasting disgrace that in an enlightened and civilized country like the United States there should be no transcontinental highway connecting the states of the two

seaboards. Even over the route selected, which I believe to be the best that has been found, many of the roads to be followed are no better than trails, and in some cases they are not even that. These conditions will be called more forcibly to the minds of American citizens and American legislators by the ocean-to-ocean contest than by any other means conceivable.[21]

Because of the MCA's objections, organizers of the New York–Seattle race tightened the rules. (See Table 3.1.) Originally, each team was allowed to change such parts as engines, gears and axles—not once but twice, if necessary. Organizers later forbid the replacing of such parts. To enforce this rule, the technical committee of the Automobile Club of America stamped the engine crankcase, cylinders, transmission case, steering gear, frame and both axles of each car. (See Fig. 3.2.) Officials then "took accurate measurements of where the stamps were placed and their exact nature," *Motor Age* said. "Wax impressions of these stamps were forwarded to Seattle so that the examiners could ascertain absolutely if the parts bore the genuine stamps they received in New York."[22]

Table 3.1 Rules of the New York-Seattle Race*

- At specified checking stations, judges will sign each auto's "passport" and crewmen will sign and mail "a prepared card" to the referee.
- Teams arriving late at any of the six 12-hour overnight stops between New York City and St. Louis must rest 12 hours, even if it means falling behind the other contestants.
- Adding crew members after the start from New York City is prohibited.
- Contestants will travel without official observers but parts will be stamped at the start. At the finish, each auto must carry the original stamps on its frame, axles, crankcase and cylinders, transmission case and steering gear. "All other parts can be changed at will." Contestants who replace stamped parts will receive a certificate of performance at the finish but are ineligible for prize money.
- Contestants may not travel on railroad tracks using flanged wheels.
- Officials will award only first and second prizes if 10 or fewer cars start from New York City. Entrants must observe local speed limits. In matters these rules fail to address, the Automobile Club of America's racing rules apply.

*Condensed and paraphrased from the Automobile Club of America's final rules.

Source: "Rules and Regulations of the International Contest for the Transcontinental Trophy," in *Transcontinental Automobile Guide: Ocean to Ocean Automobile Endurance Contest, M. Robert Guggenheim Trophy* (Seattle: J.G. Dresen, 1909), pp. 11, 13, 15.

When rule changes failed to induce other automakers to enter, Henry Ford offered a challenge to race any competitor, under either the old or new rules, for "any sum of money the acceptants may suggest as a suitable purse." He aimed the challenge at automakers or "any outsider who while not a manufacturer or possibly even an owner, does not believe in Ford cars, considers them too light, too cheap, etc.," according to *Ford Times*, the factory magazine. "One young man with sporting proclivities has taken a chance on No. 3 [the Stearns] and put up a hundred, but somehow the men we thought would welcome a chance to prove what they claim have kept silent."[23]

Fig. 3.2 An Automobile Club of America official stamps the front axle of the Stearns racer before its start from New York City. (NAHC)

The rules had little to do with the lack of entrants, the Ford Motor Company acknowledged in 1959, the 50th anniversary of the race. Rather, many members of the ALAM, the licensed group of automakers, "were literally afraid of losing their licenses if they dared enter a race that didn't have the association's approval."[24]

Thomas Blazes Trail to Denver

To spark interest in the race, organizers recruited the Thomas Flyer that won the 1908 New York–Paris race. With George Miller of the Thomas factory, a mechanic on the 1908 crew, behind the wheel and L.W. Reddington[25] aboard as the official "pathfinder," the car left New York City on March 20 to blaze a trail for the racers. (See Figs. 3.3 and 3.4.) Reddington, representing Mills & Moore of New York City, the managers of the contest, was also a "newspaper correspondent," said the *Seattle Post-Intelligencer*, with no further explanation.[26] Reddington said he had hopes of reaching Seattle by April 25.

In his push for good roads, young Guggenheim was following in the footsteps of his uncle, Simon Guggenheim, a Colorado Republican who served in the U.S. Senate from 1907 to 1913. Senator Guggenheim "has for years been endeavoring to arouse interest in a transcontinental highway, and when President Taft was Secretary of War he looked with favor upon a plan that was then being considered for a National highway," the *New York Times* said.[27] Thus

Fig. 3.3 An unidentified crewman poses with an Alaska-Yukon-Pacific pennant in the back of the pathfinding Thomas. (NAHC)

"this [New York–Seattle] route once established will mean its maintenance as a permanent transcontinental route from ocean to ocean," as *Motor Age* asserted. On their pathfinding trip, Miller and his crew would also "establish the assurance of a sufficient supply of gasoline along the route of the race to preclude the possibility of extortion on the part of dealers."[28] Accompanying Miller and Reddington were Clarence W. Eaton, mechanic, and J.S. Eley, "official photographer."[29]

Their car headed west on the route most transcontinentalists used—through central and northern New York state and along the south edge of Lake Erie to Cleveland. "At Cleveland, a modification of the original route will be made and a more southerly path tried," *Motor Age* said. "By this means the black gumbo will be avoided."[30]

Miller drove the Thomas Flyer into Chicago at 10 p.m. Tuesday, March 30, 1909. His findings? The worst roads of the trip were south of Buffalo, New York, home of the Thomas factory. The Thomas spent 10 hours traveling the 44 miles between Buffalo and Fredonia, New York, "because of the clay mud on the highways." The crew members were resting in Chicago on Wednesday, March 31, when they got word of the MCA's March 30 decision to oppose the race. Reddington "did not express himself as perturbed," the *Chicago Daily Tribune* observed.[31]

Fig. 3.4 The Thomas preparing to leave New York City on March 20, 1909. Miller is at the wheel, Reddington (wearing a moustache) at the far right in the back seat. The unidentified suited man is not a crewman. The remaining two men are evidently mechanic Clarence W. Eaton and photographer J.S. Eley, but information is lacking to match names with faces. (NAHC)

From Chicago, Miller headed southwest across Illinois to St. Louis, and then traveled due west across central Missouri and northern Kansas into Colorado. There, he angled northwest to Denver and north to Cheyenne. The Thomas pathfinder "sends back reports that unusually fine roads have been encountered in the Middle West States," the *New York Times* said, "and Pathfinder Reddington says that a National highway could be constructed with far less trouble and expense than is generally believed." In Kansas, "enthusiastic receptions were met, automobilists and farmers joining in the celebrations," *Automobile* said.[32]

Money for Highways

From Cheyenne, the racers would travel across southern Wyoming to Granger, where most drivers on an east-to-west transcontinental trip would turn southwest for Utah. But not the New York–Seattle contestants. From Granger, "the route has again been changed in order to take advantage of the recent action of the Washington state legislature, which has appropriated $120,000 toward making and repairing the roads in eastern Washington over which the contestants will travel," *Motor Age* said.[33] To reach Washington, the cars would travel through southern Idaho and northeastern Oregon. Pennsylvania, as well, had appropriated

$300,000 to begin constructing a highway from Philadelphia to the Ohio line, added the *New York Times*, and

> Gov. Brady of Idaho has secured the appropriation of $50,000 for the improvement of the roads in his State over which the ocean-to-ocean contest will be run.... Those who are interested in the good roads movement say that it would not be difficult to secure appropriations from the States of Missouri, Kansas, Colorado, and Wyoming.[34]

Guggenheim evidently agreed, according to a *Motor Age* article from early May 1909. "Robert Guggenheim, of the wealthy Guggenheim family, will announce in a few days completed plans for the incorporation of a $100,000,000 company to build a national highway from New York to Seattle. John D. Rockefeller and J.P. Morgan are interested with the Guggenheim family in this, it is said." Far to the south of Seattle, San Francisco officials begged to be placed on the route, offering to double the prize money and mint a second trophy. More direct, "Moberly automobilists actually kidnapped the pathfinders," *Automobile* reported, though that Missouri town, like San Francisco, was nonetheless excluded from the race route.[35]

"Stray Missile" Hits Thomas

On Wednesday, April 14, the trailblazing Thomas drove from Denver to Cheyenne, where it began a 12-day grind across Wyoming. The pathfinder "is battling with snow, deep mud, broken bridges and no roads at all," *Automobile* reported. Miller and his crew on Thursday, April 15, pushed through snow and mud—sometimes literally (see Fig. 3.5)—to travel the approximately 50 miles from Cheyenne to Laramie. They waited in Rawlins Sunday, April 18, "while a gang of laborers by the order of the road commissioner tried to make some of the roads west passable." Said Reddington: "The county commissioners set a crew of fifty men, with twenty teams, at work and within a short time built three culverts and two bridges and graded the road so that we could pass."[36]

On Wednesday, April 21, "the car was dug out from a mudhole near Bitter Creek by a railroad section gang of fifteen men and then placed on a corduroy road specially built for it," *Automobile* said.[37] Upon reaching Seattle, Reddington related the western Wyoming incident in more detail:

> Small lakes were encountered where we expected to find roads, and we were compelled to make detours of five to ten miles to avoid sticking the machine....

> Three miles east of Bitter Creek, we skated into a sink hole that was covered with a crust of clay and stuck fast. We got six horses, but they could not budge the car. Finally the Union Pacific sent us a section gang of sixteen men, and after working eight hours, the car was pulled out on a corduroy bridge [of logs laid on the ground].

Fig. 3.5 Miller steers while Reddington pushes just west of Cheyenne, Wyoming. (WDCR)

> After that we stopped frequently to cut sage brush to pile into mudholes to save a second experience of the sort. We carried an ax, a shovel and block and tackle, and we made good use of them frequently.[38]

The Thomas Flyer, a "scarred veteran," arrived in Seattle sporting a bullet hole in its body, which "told the story of a stray missile fired at a coyote by a Wyoming sheep herder," said the *Seattle Post-Intelligencer*. "That the bullet did not hit one of the occupants of the car is only one of the evidences of the good luck that accompanied the pathfinders."[39]

Sunday night, April 25, "was spent with a camp of sheepherders out on the plains, with the temperature at zero." The Thomas reached Montpelier, Idaho, on Monday, and Pocatello on Tuesday, April 27, *Automobile* said.[40]

Trailblazers Reach Seattle

But the pathfinding auto had a potentially serious accident on April 29, evidently in eastern Idaho. "While bucking a supposed snowdrift it went over a hidden precipice, falling some 50 feet down the mountain side into a gully," *Motor Age* reported. Miller, who was driving, walked three miles to a Union Pacific section house and returned with a gang of laborers and "extra long tackle" to haul out the car.

Though the crew apparently escaped uninjured, the Thomas had a badly bent front axle and steering arm. With the damaged parts in tow, Miller and Eaton left the other two crewmen with the auto and rode a track-inspection car back to the section house to make repairs. "After working 8 hours, Miller and Eaton, with the help of no other tools than a sledge hammer and a coke stove, straightened out the parts and proceeded back to the Thomas."[41] (See Fig. 3.6.)

In an interview with the *Seattle Post-Intelligencer* at the end of the ocean-to-ocean journey, Reddington related what was possibly the same incident, though his account and *Motor Age's* differed in nearly every detail. One reason the car was nearly a month late arriving in Seattle, Reddington said, was

> the fact that we drove off a small bridge in Idaho. The drop off the end of the bridge was not more than two feet and a half, but the shaft connecting the steering sector with the arm was snapped and it took three separate trips to Pocatello, a distance of 100 miles, before we could get a job that would stand.
>
> The accident delayed us a week. When we got out of Pocatello and were running between American Falls and Twin Falls the car hit the only stone in the road and snapped that shaft again.
>
> It was at this point that we had what could be called the only real hardship of the journey. The car was broken down in the desert, fifteen miles from the nearest ranch

Fig. 3.6 The shift lever and linkage resting on the hood and the parts strewn on the ground suggest that the Thomas crewmen are making transmission repairs. The location is undisclosed and the incident went unreported in press accounts of the pathfinding journey. (NAM)

house. The mechanic, Mr. Eaton, and I took the broken part of the machine and walked to the ranch, leaving the other two with the car. We promised to send back food and water to the men in the car, but it was eighteen hours before they got anything to eat or drink. Had the situation been serious they could have tapped the radiator but they waited and were rescued.

After reaching the ranch house we drove twelve miles to the nearest station and then rode on the train sixty miles to Twin Falls, where we found a machine shop.[42]

George Schuster, a Thomas crewman during the car's 1908 New York–Paris race, recalled erroneously that "the car broke down in Idaho and was quietly returned to the factory."[43] The first part was true; the second part was not, as the Thomas received a rousing welcome when it pulled up beside the Alaska Monument on the grounds of the Alaska-Yukon-Pacific Exposition at 3 p.m. on Wednesday, May 19, 1909. (See Fig. 3.7.)

"More than fifty automobiles from Seattle met the pathfinder at Kent yesterday," said the *Seattle Post-Intelligencer* of Thursday, May 20, "and the procession into the city came on the

Fig. 3.7 The Thomas car (background) arrives on the exposition grounds in Seattle. Miller is at the wheel, Reddington leans over the front seat. (UWSC Nowell X1200)

first speed in a swirl of dust. At the exposition grounds nearly 1,000 persons had gathered to greet the pioneers of the automobile route to the Northwest."[44]

Their intended route over the Cascades was through Snoqualmie Pass, which included a five-mile ferry ride. But "we were informed that the snow was from six to eighteen feet deep in the pass and that the ferry boats on Lake Keechelus had not been taken from the winter boat houses,"[45] Reddington told the newspaper. Thus the pathfinding crew shipped the Thomas Flyer over the Cascades by rail, from Easton on the east slope to Ravensdale on the west—perhaps 40 miles. (See Fig. 3.8.) The Flyer's odometer registered 36,979 miles, which included its round-the-world trip, mileage added since then, and all but 40 miles of the 4,001-mile trek from New York City to Seattle.

"Later in the summer there will be no difficulty in negotiating Snoqualmie pass," Reddington told the *Seattle Post-Intelligencer*. "From Easton to Ravensdale was the only time that the pathfinder was off its own wheels. We arrived at Ravensdale at midnight last night [May 18] and started for Seattle at 6 o'clock this morning."[46]

How Easy to Build?

In April, Reddington had said it would be relatively easy to build a transcontinental highway. The colorful, highly imaginative, telegrams he sent back East suggested otherwise. The *New York Times* of May 23, 1909, summarized Reddington's accounts:

Fig. 3.8 To avoid deep snow in the Cascades, crewmen load the pathfinder onto a railcar at Easton, Washington. Reddington holds a camera; the other men cannot be identified. (NAM)

> The Thomas car has slid over precipices, down mountain sides, and has been stuck in quagmires, according to Redington [sic]. It has also been drenched by cloudbursts, whirled about by cyclones, smothered by avalanches, buried by sand storms, captured by Indians, chased by wild animals, and lost in the desert.
>
> The Thomas car arrived in Seattle a few days ago and was welcomed by hundreds of motorists. The car blazed the trail across the continent as far as the Cascade Mountains. Snoqualmie Pass was to have been negotiated, but the pathfinders learned that there was still ten feet of snow covering the road.
>
> Working crews have been sent out from Seattle to clear the pass before the latter part of June or early in July, when the contestants will arrive. There are many fallen trees to be chopped away and boulders have rolled down the mountain on to the roadway.[47]

In light of such obstacles as Snoqualmie Pass, race promoters hoped the U.S. military would help build the highway between New York City and Seattle. "It is expected that the War Department will become interested enough in the present movement to appoint an observer to take the trip across the continent in the Guggenheim Trophy contest, and also to appoint a committee of engineers to look into the matter and report on the cost of such a project,"[48] the *New York Times* said. Years later, in 1942, U.S. Army engineers blazed the 1,636-mile Alaska Highway through "unbroken wilderness,"[49] but press accounts make no mention of a War Department representative taking part in the 1909 New York–Seattle race or in later efforts to improve that route.

Reddington was reportedly planning to prepare a route guide in time for the start of the New York-Seattle race. Though Reddington is not credited as its author, his observations were recorded in *Transcontinental Automobile Guide: Ocean to Ocean Automobile Endurance Contest, M. Robert Guggenheim Trophy*, a 64-page, 50-cent booklet published in Seattle in 1909 by J.G. Dresen. Eley reportedly took 1,000 photos during the pathfinding trip.[50] None of his images found their way into the Dresen booklet, evidently because it was assembled hurriedly after Reddington reached Seattle.

"Bitter Opposition" Thins Starters

"Entries comprising the principal makes of France, Italy, Germany, Great Britain, Belgium, and the United States are expected," the *New York Times* said on March 1, 1909.[51] Early on, Guggenheim spoke hopefully of 20 cars entering, including some foreign makes. Thirteen cars were entered as of the May 15 deadline. Just five started on June 1. But "the contest has met with bitter opposition from the Manufacturers' Contest Association," the *New York Times* noted, "and Mr. Guggenheim feels that he has gained a great victory in getting five cars to start in the face of such powerful antagonism."[52]

In fact, Guggenheim—"waving the white flag of truce in one hand and a distress signal in the other," as *Motor World* put it—"was in a fever of activity" during the week preceding the race,

152

attempting to get more late entries. First, he met with officers of the ALAM, the American Motor Car Manufacturers Association and the Importers' Salon, *Motor World* recounted. Then,

> he dispatched Tom Moore to Buffalo, to see Frank B. Hower, the chairman of the con-
> test committee of the American Automobile Association, to learn what terms could be
> made. Hower demanded an unconditional surrender, the turning over of all entrance
> fees, the running of the contest under A.A.A. rules and $5,000 cash for either himself or
> the A.A.A. to take hold of it. The last item quickly sent Moore to the railroad station
> for a return ticket.[53]

Guggenheim secured two of the most prominent starters anyone could find—the mayor of New York City and the president of the United States. From the White House at 3 p.m. Tuesday, June 1, 1909, President William Howard Taft, the good-roads supporter, would touch a "golden key" to simultaneously start the auto race in New York City and open the Alaska-Yukon-Pacific Exposition in Seattle, a continent away. Guggenheim had given "a Yukon gold revolver" to Mayor George B. McClellan, who was to fire it upon receiving Taft's telegraphic signal, which would start the transcontinental race.[54]

The first spectators arrived at City Hall Square, a New York City park, as early as 2 p.m. Belching clouds of blue smoke and heralded by thunderous, unmuffled exhaust, the race cars soon rolled into the square, "arrayed in all the fittings calculated to help in crossing the plains and trails of the Western country."[55]

The cars lined up facing the main entrance to City Hall: an Acme, the sole 6-cylinder entry; an Itala; a Shawmut; and two 1910 Model T Fords, the lightest cars in the race. (See Table 3.2.) "Henry Ford was on hand to see that the little machines were started safely, assisted by his New York and Philadelphia branch managers," *Automobile* reported. But according to biographer Allan Nevins, Ford, who was in New York to testify in the Selden Patent case, greeted the Ford crews and returned to the courtroom before the start.[56] (See Fig 3.9.) Missing from the field was a sixth car, a Stearns racer entered by private owner Oscar Stolp. "A telephone message from the owner said that some trouble would prevent a start with the others," confided the *New York Herald*.[57]

"Enormous Crowd" Waves Bon Voyage

As 3 p.m. approached, 8,000 spectators were "thronging the square or viewing the proceedings from the windows of adjoining skyscrapers," the *Herald* estimated. *Automobile Topics* observed "several thousand persons" while the *Seattle Post-Intelligencer's* correspondent put the turnout at "more than 20,000 people." New York-based *Automobile* called it an "enormous crowd"—without venturing to guess how enormous.[58] A cordon of 300 police officers, many on horses, separated the cars and spectators.

Table 3.2 Cars and Crews in the 1909 New York-Seattle Race

Make/entry number:	Ford No. 1	Ford No. 2	Acme (4)	Shawmut (5)	Itala (6)
Crew[1]	H.B. Harper Frank Kulick	B.W. (Bert) Scott C.J. (Jimmy) Smith	James A. Hemstreet Jerry Price George Salzman Fay R. Sheets	E.H. (Earle) Chapin R.H. (Robert) Messer T. Arthur Pettengill	Elbert Bellows Gus Lechleitner (New York City-Chicago) Egbert Lillie (St. Louis-Cheyenne) F.B. Whitmore
Body Style	Model T runabout with racing body	Model T runabout with racing body	Model 21 runabout	–	–
Weight (lbs.)	950	950	3100	4500	ca. 4000
Cylinders	4	4	6	4	4
Horsepower	20	20	48	40	45
Weight-to-Power Ratio (lbs./hp)[2]	48:1	48:1	65:1	113:1	89:1
Factory Price	$900	$900	$4,500	$4,500	[3]

[1] Frequently, press accounts referred to the drivers using a variety of first names and misspellings. For instance, C.J. Smith often appeared as "Charles" or "James," but he went by Jimmy. Pettengill's initials appear as "F.A." almost as often as "T.A." And, altered by imaginative editors or typesetters, F.B. Whitmore often emerged as "Whittemon" or "Whittemore." This listing, culled from dozens of articles, gives the standard version of each name. Some reports named crewmen by job title: driver, relief driver or mechanic. But contradictions in these accounts—one man often emerged wearing all three titles—render them worthless. Given the length of the trip and the all-night drives, it is presumed that every member of the crew drove.

[2] Dividing the car's weight by its horsepower yields the number of pounds per horsepower—a basis for comparing theoretical performance when autos vary greatly in weight and engine size.

[3] Accounts of the New York-Seattle race do not make clear whether the Itala is a 1908 or 1909 model. A 1909 40-horsepower model cost $5,250 and a 1909 50-horsepower model cost $6,250, according to *MoTor's* 1909 *Motor Car Directory*.

Sources: Accounts of the race in newspapers and auto journals; *Automobile*, "Details of the 1909 Cars," Dec. 31, 1908, p. 928, and "Essential Details of the 1909 Car List," Feb. 4, 1909, pp. 219–20; and H.B. Harper, *The Story of the Race* (Detroit: Ford Motor Company, 1909?), 32 pp.

"It was a picturesque setting for the start of the great transcontinental run," according to the *New York Times*. "A telegraph instrument had been put in place under the facade of the marble building, to receive the signal from Washington the instant President Taft touched the button there."[59] Because of his late arrival, McClellan had missed his chance to start the 1908 New York-Paris racers. This time, he arrived on the scene with moments to spare, as the *New York Herald* described it:

Shortly before the hour a big limousine brought Mr. and Mrs. Daniel Guggenheim [Robert's parents] with other members of their family, including Mr. Robert Guggenheim, the youthful donor of the trophy for which the cars were competing.

Three minutes before the time set the Mayor made his appearance on the steps of the City Hall, holding in his hand the gold mounted revolver with which the signal was to be given. He at once became the target for a battery of photographers, an ordeal that he faced with much good nature. He consulted his watch frequently, and as the seconds ticked off the approach of the starting time raised the weapon in the air.

Fig. 3.9 The Fords appeared in City Hall Square with three other entrants to await the start of the race. Scott grasps the wheel of Ford No. 2, flanked by Smith. Visible directly behind the car's radiator cap is Henry Ford, wearing a black hat. (LOC LC-USZ62-33732)

Immediately there came a roar from the engines of the five cars and a great cloud of exhaust gas raised as they awaited the shot. When it came the police car, carrying Lieutenant O'Rourke, who was charged with seeing the field safely out of the city, got under way, and the contesting cars followed quickly, to the accompaniment of a roar of cheers from the crowd.

Up Broadway and until well into the suburbs of the metropolis the cars were awaited by thousands of interested spectators, and at all points they received an ovation.[60]

Fords Stripped for Hard Service

The lieutenant's police car was actually "a conspicuous yellow taxicab [that] proceeded up Broadway, through 'Automobile Row' and on toward Poughkeepsie," *Automobile* said, adding:

> The Fords attracted a great deal of attention, not only because of their size but also because of their general get-up. The motors had no mufflers, exhausting through an opening in the side of the hood into the air. The body consisted of two seats for the drivers, with a platform at the rear for carrying baggage and supplies, covered with a brown canvas.

The chassis was stripped of everything not absolutely necessary, such as fenders[,] and the running gear was painted a dull gray. Natural colors remained in all other parts, such as aluminum unfinished for the bodies, sheet iron unpainted hoods, and unpainted gas and oil lamps.[61]

Newspapers along the route joined the *Syracuse (N.Y.) Post-Standard* in noting the size differential among racers: "Of the cars entered in the big race the most likely looking ones are the Shawmut and the Acme, with the Itala showing up as some racer, too. The first two are built on long rangy lines with large power plants, big wheels and stripped chassis.... The two little Fords look like pygmies beside the other cars as they snort along through the dust, keeping up their end of the game with apparent relish."[62]

In a 32-page post-race publicity booklet, *The Story of the Race*, the Ford Motor Company contends its racers weighed 1,200 pounds apiece—or as much as a 1910 Model T touring car. Their actual weight was 950 pounds, declared the *Chicago Daily Tribune*. On the 50th anniversary of the race, the Ford company in 1959 concurred that both racers were "stripped down to a weight of 950 pounds." Henry Ford was so weight-conscious that each racer carried a special set of Solar-brand headlamps made of aluminum instead of the customary brass. "The saving in weight in each case is eleven pounds," *Motor Field* noted.[63]

His was not a fleeting concern, as Ford Motor Company pioneered the use of vanadium steel in American autos years before the 1909 contest. The high-strength alloy allowed for the construction of Ford autos with stronger and lighter components than other automakers were using. In his 1923 autobiography, *My Life and Work*, written with Samuel Crowther, Henry Ford doesn't even mention the 1909 race. But he does declare his reasons for building the Model T, and, presumably, the point he hoped to make by winning the 1909 contest:

> I cannot imagine where the delusion that weight means strength came from. It is all well enough in a pile-driver, but why move a heavy weight if we are not going to hit anything with it? In transportation why put extra weight in a machine? Why not add it to the load that the machine is designed to carry? Fat men cannot run as fast as thin men but we build most of our vehicles as through dead-weight fat increased speed![64]

The Fords, "exciting considerable comment," carried no spares for their Firestone tires. Henry Ford told Harvey Firestone: "The car is light enough and the tires are good enough to go through without extras."[65] He might have added: "Ford agents on the route have tires enough …" because the cars would need them. Ford planned to have his local agents act as pilots, or guides, to direct the factory's racers over unfamiliar roads—quite legal under the rules of the race. "The Ford organization for piloting the little cars extends across the continent and will be a distinct advantage," the *New York Times* predicted.[66]

Frank Kulick, a well-known racer, and Ford Advertising Manager H.B. Harper, editor of *Ford Times*, the "official voice" of the Ford Motor Company, would drive Ford No. 1. "Anybody who knows the editor would never pick him out for a professional racer," Harper quipped in *Ford Times*.[67] No matter: he was traveling with one. Kulick and a co-driver made headlines in winning a Detroit 24-hour race on June 21 and 22, 1907. Alternating between two 6-cylinder Fords, they drove a total of 1,135 miles—a world's record for 24 hours had they used a single auto.[68]

Taking charge of Ford No. 2 were Ford employees Bert Scott and C.J. "Jimmy" Smith.[69] Did the Ford company seek volunteers for the grueling trek? No, according to Smith in a 1951 interview: "I don't know how the other drivers were picked for that. They came and told me that Bert Scott was the driver of the car and I was to be the mechanic."[70]

Other Three Are Ex-Race Cars

The yellow Acme Model 21 runabout[71] "is fitted with bucket seats for four people, and the big gasoline tank is between them," *Automobile* said. "Tires are carried at the rear and supplies in large boxes on the running boards. The Acme was the largest of the quintette, and carried a full load of passengers."[72]

It remains unclear who entered the Acme in the New York–Seattle race. The *New York Times* listed the entrant as the "Cordner Motor Car Company" of Reading, Pennsylvania.[73] But Cordner was an Acme agent in New York City, where the racer was put on display for a week before the start. Reading was home to the factory, which the May 20, 1909, *Automobile*, says entered the auto. At the start, in fact, the car carried a special manufacturer's license plate— but one issued in New Jersey, not Pennsylvania.

The Acme crew included James A. Hemstreet, who in 1916, as an American Automobile Association representative, would officiate when a Hudson auto would make a record-setting transcontinental round trip; and George Salzman, the famed Thomas driver. (See Fig. 3.10.) The Acme was the same one "that finished second in the Fairmount Park race last October, and which won the Campbell Trophy for consistent showing," the *New York Times* said. At Philadelphia's Fairmount Park, George Robertson's Locomobile won the October 10, 1908, 200-mile Founder's Day Cup Race, ahead of L. Patchke in an Acme and Ralph Mulford in a Lozier. A broken crankshaft sidelined Salzman's Thomas, one of 16 starters.[74]

The Itala "was entered at a late moment, but it was fully prepared, with a special demi-ton-neau body for four passengers and with the supplies carried in the tonneau and the tires at the rear. It is painted a lead color and stripped down as much as possible," *Automobile* noted. The same car "won third place in the Motor Sweepstake races on Long Island last fall," *Motor Field* said. "The only changes made for the transcontinental race was the addition of an extra large [gas] tank and four seats with the necessary equipment for the journey."[75] (See Fig. 3.11.) The Itala Import Company of New York City entered the auto, according to *Motor Field*. Actu-ally, countered *Motor World*, Guggenheim entered the car himself:

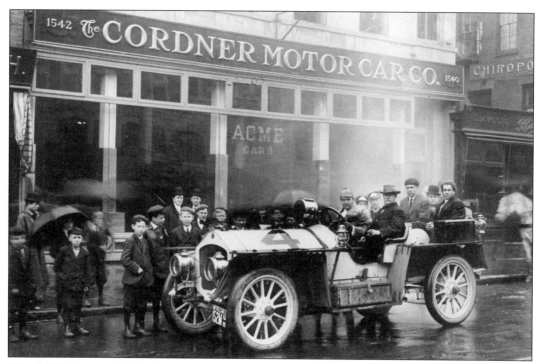

Fig. 3.10 It is most likely George Salzman posing at the wheel of the Acme, on which canvas and light wood replaced the standard metal fenders. The other men are unidentified. (NAHC)

> According to the Itala press agent, he [Guggenheim] was so impressed with its ability to start on short notice, that he "could not resist its purchase." The fact that the car was needed to give the affair an "international" complexion is not dwelt upon.[76]

The white Shawmut carried "extra large wheels, using 40-inch wheels and tires, with a body for four men, tires and supplies being carried between the two seats. These are extra high at the back and have small side strips carefully upholstered to act as head-rests when one set of drivers takes a sleep," *Automobile* said. (See Figs. 3.12 and 3.13.) "The Shawmut and Acme are the only ones carrying running boards and mud-guards."[77]

The transcontinental Shawmut "is the identical chassis which did battle in the [Briarcliff] road race last spring, and which last fall tied for first place in the endurance run of the Bay State Automobile Association, traveling over 1600 miles without a single mishap or adjustment," according to the *Stoneham (Mass.) Independent*. "The car has the identical mechanical parts which went into it when originally built, no defect developing in two seasons of road work."[78]

One of 22 entrants in the 240-mile race over a 32-mile road course around Briarcliff Manor, New York, on April 24, 1908, the Shawmut "ran perfectly from start to finish" and placed 11th, according to a Shawmut factory catalog. All the cars that beat the 40-horsepower Shawmut had at least 15 more horsepower, the factory claimed.

Fig. 3.11 The Itala and its unidentified crew. Visible on the back of the car is the large gas tank added for the race. (NAHC)

Harold Church drove the car to a first-place tie with a Franklin and a Studebaker in the Bay State Endurance Run of 1908, which ended October 1. Each of the three cars ran 1,607 miles and survived a runoff that ended when the automakers agreed to withdraw their autos, "as the cars were in first-class shape and apparently could keep on indefinitely."[79]

Truly a survivor, the Shawmut was one of two autos that had been saved from a November 13, 1908, fire at the Shawmut Motor Company in Stoneham, Massachusetts, the *New York Times* said. The company moved to nearby Boston but had evidently built its last

Fig. 3.12 The Shawmut entered the race equipped with canvas fenders, an ax strapped to the front seat and a foot locker replacing the left running board. From left to right: T. Arthur Pettengill (at wheel), Earle Chapin and Robert Messer. The pennant reads "Boston Motor Club." (AAMA)

autos. "Thus it was that the Shawmut went into the contest without the support and aid of a factory and carrying only a few spare parts."[80]

Driver Arrested at Erie

John H. Gerrie, representing the Automobile Club of America, would act as a "pacemaker" from New York City to St. Louis. One source identified him as an automobile writer for the *New York Herald*.[81] To check their speed, Gerrie would ride ahead of the racers in a non-competing 6-cylinder Ford, which the contestants were required to follow. His role was largely ceremonial, for he allowed cars to pass him and to speed, as well. Keeping together, the racers spent the first night, Tuesday, June 1, in Poughkeepsie, the second in Syracuse, New York, and were scheduled to spend the next four nights in Buffalo; Toledo, Ohio; Chicago; and St. Louis. The rules required the contestants to lay over 12 hours at each of these six "night controls."

The drivers showed an early propensity to get lost in the wilds of New York state. Near Peekskill on the first day's 73-mile run, "the Acme lost the lid of its tool chest, and in stopping for it got off the course, traveling ten additional miles before finding the route. The Ford, driven by Bert Scott, also got off the route, and arrived at Poughkeepsie with the Acme, twenty minutes late," according to the *New York Times*. But the *Poughkeepsie Daily Eagle* reported that the Shawmut and two Fords arrived at Nelson House, a hotel checking station, at 7:06 p.m., followed by the Itala at 7:12 p.m. and the Acme at 7:26 p.m. "The men were covered with dust and there was a good sized crowd in front of the hotel to watch the cars come in."[82]

On Wednesday, Day 2, the five cars drove 207 miles to Syracuse, where the men spent the night at the Yates Hotel, noted the *Syracuse Post-Standard*. "Not an accident has yet occurred to mar the race and the drivers and mechanicians are all elated over the prospects of a great race." On Thursday, contestants covered the 150 miles to Buffalo within the allotted time. "At the Bison City," observed the *Chicago Daily Tribune*, "all five cars were on even terms."[83]

Fig. 3.13 A rear view of the Shawmut shows Messer leaning on the side and Chapin standing in the back seat. (NAHC)

That would change on Friday, Day 4, when the contestants

were allowed 15 hours to travel the 296-mile route to Toledo, passing through Erie, Pennsylvania, and Cleveland. Leaving Buffalo at 5 a.m., the cars drove through "a constant downpour of rain," according to the *Cleveland Leader*. "Tire troubles were numerous and during the storm the Acme and Shawmut cars skidded into ditches, the former being hauled out by a trolley car," according to the *New York Times*.[84] Offering more details, *Automobile Topics* said the Acme slid into a deep ditch and missed plunging into a swamp

> only by striking a tree at the foot of the embankment. All the crew except George Salzman jumped as the car left the road, but none were injured. After four horses had occupied four hours trying to pull the car out of the ditch a trolley car came to the rescue and drew the Acme up on the road again. The only damage sustained to the car was smashed head lamps. The Shawmut and Itala cars came along while the Acme was down the hill and their crews insisted on helping the Acme crew out of their trouble.[85]

The *New York Times* mentioned a rainstorm between Cleveland and Fremont, Ohio. According to the *Cleveland Leader*:

> So far the race has proven uneventful, save for the arrest of Fay R. Sheets at Erie, driver of the fifty horse power Acme car. As the car approached Central Park at Erie a policeman hailed the driver, and placed him under arrest for violating the speed laws.

> Sheets was locked up, but the timely appearance of the mayor made it possible for Sheets to continue his way three minutes after his arrest.[86]

Unchastened, Sheets "deposited $25 for his appearance in police court this afternoon, and then flew out of town as fast as he came in." Later on Day 4, as Smith of Ford No. 2 recalled in a 1951 interview, "coming into Cleveland, we were all of us going a little too fast, and we all got pinched—all of us in the group. They fined the whole bunch of us." A 1959 factory history qualified this and changed the city: "The two Fords ... came into Erie so fast that the crew of both cars were arrested for speeding and had to pay fines."[87]

A Cleveland Ford agent met the Ford racers at Erie and piloted them into Cleveland about noon. They arrived half an hour ahead of the others, "resembling piles of mud," press accounts indicate.[88] All contestants passed the Itala about 50 miles northeast of Cleveland at Geneva, Ohio, where the Italian entry evidently stopped for an unspecified reason. In a 1959 interview with the *Boston Sunday Globe*, Shawmut crewman Earle Chapin supplied the reason:

> The first leg [to St. Louis] had roads of a sort but not racing roads. The Itala driver found that out near Erie, Pa. He passed us lickety split and in a short time we saw him hung up on a tree stump with all four wheels off the ground. They had been cutting big trees right beside the road and the one the Itala straddled must have been six feet in diameter.[89]

Guided Fords Take Lead

Contrary to the *Cleveland Leader's* report, Friday's run was eventful—especially in retrospect. For Friday was the day both Fords gained a lead they held all the way into St. Louis (where the race officially began) and beyond. By checking in at Toledo's Secor Hotel at 7:20 p.m. Friday, 40 minutes ahead of schedule, the Fords became the only autos to complete the day's run within the allotted time. "The others lost the way frequently and at other times had hard battles with the mud, the result being that the Fords reached Toledo four hours ahead" of their rivals, the *Chicago Daily Tribune* reported. Gerrie was also late in arriving Friday. Thus "the pacemaking car for the most part made only its own pace," *Automobile Topics* quipped.[90]

So fast did the Fords spin into Toledo that the next day's *Toledo Daily Blade* proclaimed in an eight-column headline: "Ocean-to-Ocean Auto Contest Is Frenzied, Foolhardy Speed Terror." A *Blade* reporter traveled with Toledo Ford agent Stanley Roberts, who drove east to meet Fords No. 1 and No. 2 at Fremont:

> When the first Ford crossed the bridge over the Sandusky river, there began a race between Roberts, in the pilot car, and Kulick and Scott in the two little machines.
>
> Although a schedule was mapped out, speed limits were entirely ignored in the mad rush across Ohio. The first Ford car to leave Fremont made the trip [about 32 miles] in one hour and nineteen minutes, over roads flooded and over bumps that threatened a hundred times to hurl the drivers into the ditches.
>
> Between Fremont and Stony Ridge there were big holes filled with water and when the speeding cars hit them water flew fifty feet over adjoining fields. As every minute gained per day counts, the Fords speeded into Toledo on the high all the way. Usual slackening for bridges, culverts and bumps was ignored. It was a case of take a chance, and everybody came through without being killed.
>
> A section of the road ... is undergoing repairs, and a soft mud highway is the only one traveled. The skidding in that was something frightful, but Roberts drove the pilot through it forty miles an hour. He finally overhauled Kulick and Scott, but once again on the stone pike, Kulick went ahead and smoked into Toledo like an express train.
>
> Fremont, usually vigilant about speed laws, let them go by the board and the whole town was out to see the cars go through. When they hit the paved street, throttles were opened wide and all available gas was burned. All along the mud roads one man was kept busy ba[i]ling mud out of his driver's goggles, and the glasses finally had to be abandoned entirely.[91]

The *Blade* said the Fords arrived in Toledo at 7:15 p.m., followed by the Acme at 11 p.m., Itala at 11:15 p.m. and Shawmut at 11:30 p.m. Thus after a mandatory night stop of 12 hours, the Fords left Toledo for Chicago Saturday morning, Day 5, four hours ahead of the others. The Ford company made much of this lead; the *Automobile Topics* report, however, indicates that the Acme's four-hour struggle to free itself from a ditch, with the help of the Itala and Shawmut teams, accounted for the Ford lead.

A "Novel Way" to Advertise

The automaker boasted that its lightweight Fords outperformed the heavier racers on two scores between New York City and Cleveland—New York's hills and Ohio's mud. "Score 3 comes in the cost to run.... Our cars fill up every night—those other cars once or twice between. We are averaging 22 miles per gallon in gasoline, while at best two of the competitors are running less than ten miles per gallon."[92]

Following the automaker's lead, Ford agents also capitalized on the widespread interest in the race. Ford agencies in Chicago, New York City, Philadelphia and Toledo "have taken a very novel way to advertise the ocean-to-ocean contest," the *Ford Times* reported. These agencies painted route maps on their showroom windows and moved small cars along the map as telegrams arrived telling of the racers' progress. "This enables all who pass by the window to keep tab on the race, and it has never failed to keep an interested crowd before the window."[93] (See Figs. 3.14 and 3.15.)

Fig. 3.14 This photo of a map painted on a Ford agent's window shows reflections of the photographer, three onlookers and a horse-drawn wagon without and a shiny new Model T within. (NAHC)

The 244-mile drive through Bryan, Ohio, and the Indiana cities of Goshen, South Bend and Michigan City "usually requires two days for ordinary tourists," the *Chicago Daily News* said. With Ford agents still acting as pilots, the Fords—aided by a four-hour head start—led all the way. "The two leading automobiles in the big race from New York to Seattle, passed through Goshen at 2:40 this afternoon," Saturday's *Goshen Daily News-Times* reported. "Both passed through without stopping. Machines and drivers were covered with mud."[94] The trailing autos would reach Goshen about 6 p.m., the newspaper said.

At Goshen, the Ford racing crews met a group of guides headed by Thomas J. Hay, a Chicago Ford agent who would pilot the cars through the Windy City and as far as Springfield, Illinois, according to the *Chicago Daily Tribune*. "With Mr. Hay in charge of the piloting and guided by these experienced pathfinders it was comparatively easy to find the way into Chicago." The Fords reached South Bend at 3:35 p.m., as the *South Bend Tribune* reported, followed by the Shawmut (6:53 p.m.); Acme (7:05 p.m.), which got lost between Goshen and South Bend; and Itala (8:45 p.m.). Arriving in South Bend "tired, dusty and in some cases hungry," all the racers stopped at the Oliver Hotel, a required checkpoint. There, they signed "the required voucher for mailing to the judges," according to the *South Bend Tribune*.[95]

The Fords' comfortable lead, the newspaper commented, was "due to the Friday run, when the two Ford entries, guided by agents throughout the course of the trip, alone managed to stay on the correct road … while the remaining cars were floundering around on the wrong road near Vermilion, O." Other racers "were handicapped by being unfamiliar with the roads,

Fig. 3.15 Another view of the route the New York–Seattle racers would follow.

no one going out to pilot them as was the case with the Fords," the *Chicago Daily Tribune* added. But the other racers were getting at least some assistance, for "the Shawmut and Acme cars left for Chicago under the guidance of Walter Sibley, a local man who spent one night with the victorious Thomas when that car passed through here" during the 1908 New York–Paris race, said the *South Bend Tribune*.[96]

The Fords separated slightly on the 40-mile drive to Michigan City, where a "local Ford delegation" met the leading car, Scott and Smith's Ford No. 2, east of the city at 5:25 p.m. "There were a dozen Ford machines in line with several guests including Mayor F.C. Miller and representatives of the press," including the *Michigan City Evening News*. "The little roadster fairly burned up the road getting into Michigan City and after a short stop at the corner of Franklin and Eighth streets, the westward journey was resumed."[97] Harper reported that "there has been no accident so far unless the death of an unfortunate chicken near Michigan City and the sudden hiking of a rural district horse for a deep ditch near South Bend can be called accidents. We spent 30 minutes getting that horse out and righting the buggy."[98]

Stearns Starts, Stops, Quits

While the five racers were hurrying toward Chicago, 970 miles from New York City, Oscar Stolp on Saturday, June 5, nosed his rejuvenated No. 3 Stearns to the starting line. (See Figs. 3.16 and 3.17.) Traveling with Stolp were Robert Maxwell and Harry Holman, drivers, and Dennis Major, mechanic.[99] *Motor World* said the car was "fitted with solid tires and Stolp's system of spring suspension,"[100] for which he received a patent (No. 850,073) on April 9, 1907. Stolp's "Equalizing Suspension Lever," which mounted two per axle at $50 a pair, "saves a very large percentage of tire expense and permits the use of solid tires without lessening the comfort of the passenger or increasing the wear on the machinery," the inventor claimed.[101] (See Fig. 3.18.)

What delayed him? "A mishap to his racer," stated the *New York Times*. "Reported engine difficulty," according to Ford driver H.B. Harper. "Its engine being apart," *Automobile Topics* asserted. The car's "disputed eligibility as a contestant," the *South Bend Daily Times* offered. "Because the garage mechanic who was supposed to have it ready by May 15 had violated his agreement, and at the last minute resigned his job," another source contended.[102]

The Seattle Post-Intelligencer, however, hinted at a more sinister scenario by blaming the delay on the Stearns factory, opposed to the New York–Seattle race and, presumably, to Stolp's use of a Stearns auto. Thus the factory, "in shipping parts requisite for the journey, did not deliver the materials to Mr. Stolp."[103]

For the Stearns, Guggenheim waived the mandatory 12-hour night controls, according to the *New York Times*. The belated four-man crew left New York City at midday Saturday. "Stolp has agreed not to exceed the speed limits of the various States through which he will

Fig. 3.16 Stolp's solid-tired Stearns and its unidentified crew. (AAMA)

Fig. 3.17 Some of Stolp's supplies for the long grind to Seattle. (NAHC)

have to pass. He will drive night and day," in hopes of catching the others at St. Louis. But to do so, the *Chicago Daily News* noted wryly, "it is anticipated that he will need a pair of wings."[104]

Actually, he needed better luck—and a good machine shop. Stolp's Stearns tarried in Tarrytown, barely beyond New York's city limits, where it broke down and could not go on. "According to the information received here the Stearns broke the transmission case and had to be towed back to New York," the *New York Times* said.[105]

A train that also started from New York City Saturday had a better

Fig. 3.18 Stolp's patented Equalizing Suspension Lever. (Jan. 16, 1909, Scientific American)

chance of catching the racers. According to *Automobile*, the managers of the New York–Seattle race scheduled the train,

> which will duplicate the Wall Street Special ... overtaking the contestants at St. Louis, stopping one day. Two days will be spent in Denver, and then four in the Yellowstone Park. The automobiles will again be seen at Boise, and the train will be at Seattle for the finish, staying five days.[106]

Stricken Driver Leaves Itala

The Ford racers pulled up to the Chicago Automobile Club at 8:50 p.m. Saturday, according to the *Chicago Daily Tribune*. "The transcontinentalists did not waste much time at the clubhouse, stopping only to check in and then to scatter for a good night's sleep." The Acme and Shawmut arrived together at 2:06 a.m. Sunday, the *Tribune* said, and "at 3:50 o'clock [a.m.] the foreign built Itala rounded the corner in a dilapidated condition, due to a bent axle."[107]

Added the *Chicago Daily News*: "The Itala bent a rear axle between South Bend and Goshen and after it got here it had to go to the repair shop." After waiting in Chicago the required 12 hours, the Itala was eligible to leave for St. Louis at 3:50 p.m. Sunday. But the car was "not in condition to go out on its schedule time,"[108] and so left at 4 a.m. Monday, June 7, 19 hours behind the Fords, according to Chicago newspapers. By all accounts, the Itala's driver arrived in worse shape than his stricken steed. Said the *Chicago Daily Tribune*:

Gus Lechleitner, driver of the Itala car, traveled the last few miles with chills and fever that sent him to bed as soon as the party reached the Great Northern hotel. The other members of the crew, F.B. Whitmore, a nephew of Warren G. Demarest, the American agent for the car, and Elbert Bellows were convinced that a few hours' rest would bring their associate back to form, but physicians later told them that Lechleitner had a severe touch of la grippe, resulting from exposure and they would have to keep him under the covers for several days at least. At his urgent request the other two members of the crew agreed to drive the car ahead.[109]

The *Chicago Daily News* concurred that the grippe, or influenza, had leveled Lechleitner. Race officials, however, received a telegram saying Lechleitner "had been taken ill with diphtheria on the road between South Bend, Ind., and Chicago," according to the *New York Times*. (*Motor World* seconded the diphtheria diagnosis.) "The driver was hurried to a hospital as soon as the racing car reached Chicago," in hopes of staving off pneumonia.[110]

Bellows could not drive a car, according to the *New York Times*, which neglected to name Whitmore as a crew member. So "the services of Lillie have been secured here [New York City] as a substitute driver, and he left [Sunday] night for St. Louis, where he will take the wheel of the Itala for the rest of the journey across the continent"[111]—if the other crews consented to the change. They apparently did: an Egbert Lillie was aboard the Itala when it reached Cheyenne, according to the *Wyoming Tribune*.

The other drivers were not even consulted, *Motor World* implied in another version: "Although under the rules a substitution of drivers is not permitted, M. Robert Guggenheim, the referee, telegraphed a special dispensation, permitting a new driver to take Lechleitner's place." But *Automobile Topics* said that when "E. Lillie" arrived in St. Louis to board the Itala as a substitute driver, Gerrie notified him "that he could not act in that capacity. He then elected to go along in the car to the coast as a passenger."[112]

Sunday's Drive "A Little Too Much"

The Fords left Chicago for St. Louis at 9 a.m. Sunday, June 6, some five hours ahead of the Acme and Shawmut's scheduled departure. Sunday's trek was 283 miles, according to race organizers, though road maps give the distance over modern, paved roads at closer to 300 miles.[113] "Some place the distance to St. Louis by road as high as 370 miles" over the hard clay roads of 1909, contended the *Chicago Daily News*. Regardless of the exact mileage, it was "a little too much to crowd into one day … [because] the roads between the two cities are notoriously bad."[114] In describing the terrain and road surface between each town on the Chicago–St. Louis route, the *Transcontinental Automobile Guide* that was prepared for the 1909 race consistently used the words "rolling" and "dirt." This was all too true, Kulick and Scott discovered.

They drove their Ford racers on a southwesterly course through Illinois on a route to Bloomington, Springfield and Litchfield—the future Route 66. The Fords reached Bloomington and left for Springfield at 5:10 p.m., with the Shawmut three hours and the Acme four hours behind. The Fords reached Springfield, 200 miles from Chicago, at 10:45 p.m.—an average pace of 14.5 mph. "The slow time made was owing to the bad roads encountered between Bloomington and this city, some of them being axle deep with mud," reported Monday's *Illinois State Register* of Springfield. "The two Ford cars appeared to be in good shape and the drivers and mechanics in the machines were in good spirits, despite the hard rough roads they had encountered," added Springfield's *Illinois State Journal*.[115] As the *Register* described it, the Fords

> stopped at the Leland hotel for lunch, and proceeded at 11:15 [p.m.] for St. Louis....
>
> Reports from St. Louis and also from the driver of the pilot car, which is a Ford, driven by H.C. Apgar of that city, have it that the roads between here and the Missouri metropolis are almost impassable. The machines showed the effect of the roads, even the seats being covered with mud. The machinery, as it was all covered, did not suffer from the going but the tire[s] and wheels were fast going "to the bad"....
>
> Mr. Scott said last night: "We have had a hard race so far, but have made excellent time considering the roads. The motors have stood the strain fine and the tires have held up well. We are well in the lead and unless we have awful hard luck, should be the first to arrive in Seattle."[116]

Leaders Reach St. Louis

The Fords drove all night to reach the Hotel Jefferson in St. Louis at 7 a.m. Monday, June 7. The Shawmut arrived about two hours later and the Acme, which "traveled fifty unnecessary miles [Sunday] when it got off the course," arrived before noon. "All the drivers report having encountered bad roads in the run from Chicago, the heavier cars being handicapped by the mud," according to press accounts.[117]

While the leading drivers turned in to sleep in St. Louis, the repaired Itala had just begun to wallow in Illinois mud. Leaving Lechleitner to recuperate, Bellows and Whitmore left Chicago at about 4 a.m. Monday. They stopped for 10 minutes in Springfield at 7:10 p.m. Monday, the following day's *Illinois State Register* reported:

> Harry T. Loper accompanied Whitmore and Bellows to St. Louis as a guide, the two men having found Illinois roads in such poor condition as to necessitate the services of some one who had been over the route before.
>
> The "Itala" came through Bloomington yesterday afternoon. Bellows took the wheel out of Bloomington, but he was compelled to give it up when about half the journey between Bloomington and Springfield had been covered. Whitmore drove the car out of Springfield.[118]

If Bellows had actually never driven an auto, as the *New York Times* contended, he was learning under battle conditions. But just 45 miles south of Springfield, as *Horseless Age* reported it, "the Itala car is reported to have met with an accident at Litchfield, Ill., having broken one of its wheels and its steering gear."[119] A St. Louis wire story on Tuesday added these details: "The Italian car, the missing fifth competitor in the New York to Seattle race, is lying on its side near Litchfield, Ill., with a wheel gone. This information was brought to St. Louis when the car's drivers came in this morning, seeking repairs and to consult with the managers of the race."[120]

At Buffalo, pacemaker John H. Gerrie left his 6-cylinder Ford pace car. Between Buffalo and St. Louis, the *New York Times* reported, he had taken turns riding in the Acme or Ford racers—meaning he was not always traveling at the head of the pack. The Acme had thus carried up to six people, including Gerrie and local pilots. The Shawmut normally carried three people and the Ford runabouts two apiece. "Over the sand and clay roads of Ohio and Indiana, and the clay trails of Illinois the big cars pounded along at a speed ranging from twenty to forty miles an hour without much effort," Gerrie said. Both states at the time restricted autos on the open road to 20 mph. "The little Fords, piloted for the entire race by relays of Ford agents, stood up equally well, and reported first at many of the night controls."[121]

"Spirited Animals" Speed West

At St. Louis, Gerrie left the contestants. "From now to the end of the contest, with all restraint thrown off, they will travel daily as far and as fast as they like," the *New York Times* said. It appeared the racers had a tacit agreement to start west on Tuesday morning, June 8, but "the Ford crews suddenly changed their plans this evening and left for Centralia at 8:15," according to a Monday, June 7, dispatch from St. Louis. "The Acme and Shawmut crews were then called out of bed and soon started after the Ford cars. The Fords, by their sudden departure, were attempting "to steal out of St. Louis without the Acme and Shawmut drivers' knowledge," *Automobile Topics* charged. The racing cars, risking a nighttime drive on Missouri's muddy roads, were like "four spirited animals which have been held in leash and then suddenly turned loose," *Automobile* observed.[122]

The drivers—now free to follow the route of their choosing across central Missouri, as long as they checked in at Centralia—generally followed a route some 20 miles north of what is now Interstate 70. The Acme arrived first in Centralia but the Fords later regained the lead. This brought the racers to Glasgow, about two-thirds into the 265-mile trip to Kansas City. Late in the race, the *Seattle Post-Intelligencer* published a detailed account of what happened at Glasgow during the dash across Missouri overnight Monday and on Tuesday, June 8:

> The roads were good—for Missouri. Centralia was the first control [checking station]. The Fords had been first to reach each control west of Erie, and the big cars were keen to get even. On good roads a high-powered racer naturally is fast, and first to reach Centralia,

150 miles from St. Louis, was the Acme, checking in at 6 a.m. At 6:10, only ten minutes later, the two Ford cars came whizzing in.

To be second and third was galling. But an additional reason existed for winning the next lap. Eighty miles west of Centralia is the city of Glasgow, on the Missouri river. Here the cars were scheduled to cross. The only means provided was a small gasoline ferryboat with a limited capacity. The first cars to arrive got the boat; the next ones would be compelled to wait until later, for the boat was old and not to be depended upon.

All Glasgow was waiting anxiously, for news had been phoned ahead. The ferry dock was crowded. Henry Ford, who was also there, was probably the most anxious spectator of all. Just at 9 o'clock [a.m.] in came the two Fords—first and second—and where they had passed the other cars was unknown. But they were there, they were on the ferry and were nearly across before any other cars came in. [See Fig. 3.19.]

The ferry engine was hitting on two cylinders and drifting down the river. Could the ferryman make a landing? He missed it a hundred yards, missed it again, then made it, and his old engine quit for the day.

Fifteen miles farther is the Boonville ferry. One of the other cars hustled there and crossed while the other took to the ties and crossed on the railroad bridge.[123]

A *Boston Herald* photo reveals that the Shawmut crossed the river on the railroad bridge at Glasgow—a dangerous stunt—and the trailing Itala did likewise. (See Fig. 3.20.) The Acme

Fig. 3.19 The two Fords on the Glasgow ferry, which is evidently unloading the racers on the west bank of the Missouri River. (HFM 0.2575)

Fig. 3.20 Over planks and discarded railroad ties, the Shawmut crosses a ditch to reach the railroad tracks at Glasgow, Missouri. (SHS)

evidently crossed on the Boonville ferry. In April, when the Thomas Flyer reached Glasgow, high winds and rough water had sidelined the ferry. George Miller looked at "the open railroad bridge, 100 feet above the river ... but the wind was so high that it was deemed best not to attempt the bumping across."[124] Instead, the pathfinding crew waited until the wind abated and the ferry ventured out again. (See Fig. 3.21.)

When the Shawmut crew arrived and learned that the ferry would spend two days on the west bank of the river for repairs, "Chapin and Messer hired a rowboat to find out what was causing the delay and they were told that the pitch of the propeller had to be changed," according to the *Stoneham Independent*. "As Chapin remarked, it might have been so, but it was very strange that the propeller on the old tub had to be adjusted at that particular time after it had been plowing back and forth for a couple of decades without trouble."[125]

Acme Hits Tree, Farmers Hit Roof

Other accounts reveal that the Fords took the lead when an accident halted the Acme. Just west of Centralia, where the Acme led, "the imp of the perverse took charge of their affairs ...

Fig. 3.21 The Thomas pathfinders briefly contemplated crossing the Missouri River on the long, high railroad bridge in the background. They waited for the ferry. (NAM)

[when] a tree on a hill side was run into during the night, throwing the alignment of the frame out. Six hundred miles further, this caused the jackshaft to break, necessitating a delay of two days and a half," according to the *Seattle Post-Intelligencer*. The Show-Me State held other adventures. Harper later told the *Denver Republican* "that while in Missouri they had been held up by angry farmers with guns and threatened with death unless they went more slowly."[126] The newspaper provided no further details and Harper neglected to mention the incident in *Story of the Race*.

The Fords arrived together in Kansas City at 5:35 p.m. Tuesday, or 21 hours, 20 minutes after leaving St. Louis. The Acme and Shawmut arrived in Kansas City about an hour later, the *New York Times* reported. Newspaper accounts don't indicate whether the crews drove through the night or stopped for sleep. The two leaders actually pulled up to the Baltimore Hotel in Kansas City, Missouri, at 5:15 p.m. Tuesday, according to a local newspaper, which quoted Harper's impressions: "The roads were just about medium between here and St. Louis. We did not have any trouble whatever until we ran onto a nail just a few miles outside of this city. The only really bad roads we have encountered since we started were in Illinois, and they were the worst I have ever seen."[127]

In *Story of the Race*, Harper says roads on this stretch "ran from bad to medium, for there had been a great deal of rain for some weeks previous."[128] Actually, both Ford autos were "battling the mud" east of Kansas City (see Fig. 3.22) recalls a Missourian who, at age 16, watched the cars crawl by her farm near Mayview:

Fig. 3.22 While two unidentified men puff for pleasure, Smith, far right, dolefully considers the mired Ford No. 2. According to one account, the scene is just east of Kansas City, Missouri. (MHI)

> Father talked with them as they plowed along. They went for about a quarter mile and then had to get out and dig the mud from the wheels. At places the cars sank into the mud clear up to the hubs.... When this happened several men would take hold of the car and simply lift it out of the mud.... The two Fords were traveling as near together as possible and in addition to the drivers, they had extra men to work on their cars.[129]

Using his own St. Louis departure and Kansas City arrival times, Harper claimed the Fords covered the distance in 20 hours, 40 minutes, "a new record for the run across the state."[130] But some enterprising Missourian, under better weather and road conditions, had surely exceeded the Fords' tedious pace.

Ford Crashes; Shawmut Takes Lead

The Acme, Ford and Shawmut drivers cranked up their engines and continued west Tuesday night, some in hopes of traveling through Lawrence, Kansas, to Topeka, 75 miles. But at about 11 p.m. "a heavy storm arose causing the crew of Ford No. 1 to cease operations until morning," according to Harper in Ford No. 1. Smith and Scott in Ford No. 2 continued, however, and had proceeded but a few miles out of Kansas City when an accident befell it which for a time seemed likely to put the car out of the race.... Driver Scott in Ford No. 2 decided to push ahead and make Topeka. In the slippery mud the car skidded down a 14 foot embankment into a stream, throwing Scott and Smith into the water.

There the car lay until [Wednesday] morning when the other Ford came along, helped pull Number 2 back on the road and then proceeded. The axle was badly bent, for the drop had been sudden. Taking it out, Scott and Smith walked three miles to a blacksmith shop, straightened the axle, returned, replaced it and started again.[131]

According to a race chronology that Ford compiled in 1959, the accident actually occurred "just outside Williamstown," about 40 miles west of Kansas City. "We went off the end of a bridge one night, down into a stream, and bent our front axle," as Smith recalled the incident in a 1951 interview. The next morning, "We found we were right near a railroad. We got the section gang there and straightened the axle out. We got it back in, and away we went."[132]

The Shawmut arrived first in Topeka at 8:15 a.m. Wednesday, June 9, with Ford No. 1 in second place at 8:45 a.m., according to a wire report from the state capital. The Shawmut crew left Topeka for Manhattan, the "Little Apple" of Kansas, at 9 a.m., a half-hour ahead of Ford No. 1. "Ford car No. 2 reached Topeka at noon today after being marooned in the mud for several hours between Topeka and Lawrence," an apparent reference to the third-place car's overnight accident. "Only a short stop was made and the car left for Manhattan, forty miles west, the next control station." The fourth-place Acme, meanwhile, was still 30 miles east in Lawrence, where it arrived during the noon hour. "The driver passed the night at a farm house and overslept, getting a late start," explained a wire report.[133]

Plowing through Kansas

The Shawmut perhaps got stuck or broke down—press accounts don't say—but it faltered west of Topeka. Thus Ford No. 1 led the way into Manhattan Wednesday afternoon. "Half an hour after the Ford racer arrived the Shawmut, No. 5 in the race, came whirling in, and was soon followed by the No. 2 Ford. The Acme arrived an hour after the second Ford," according to a dispatch from Manhattan. Offering a new version of its overnight accident, this report said that during a lightning storm Ford No. 2 crashed into a 10-foot ditch—not over a 14-foot stream bank. Regardless, the car "was in bad condition when it arrived here." It required some unspecified repairs, and left Manhattan trailing the Ford No. 1, Shawmut and Acme.[134]

Ford No. 1 sped through Salina, 119 miles west of Topeka, early Wednesday evening, 10 minutes ahead of the Shawmut. While the leaders pushed into central Kansas, the Itala "arrived here today," according to a Wednesday dispatch from St. Louis, "and will probably start West tonight."[135]

Late the next day, Thursday, June 10, the four leading cars rolled into Ellis in west-central Kansas, with the Acme in first place. "The Acme car did not stop, but the two Ford cars were delayed forty-five minutes for repairs," said a wire report from Ellis. "They got away an hour behind the Acme. The Shawmut car was twenty minutes behind the Acme, having left while the Ford cars were repairing."[136] From Ellis, most of the cars followed the northwesterly route through Oakley, Colby, Goodland and Kanorado, Kansas, and on to Burlington, Limon, Byers and Denver, Colorado.

The Shawmut was apparently in the lead when it headed north from Oakley toward Colby Thursday night, according to a dispatch from Colby. But the Shawmut "was compelled to tie up at a farmhouse, eight miles south of here, for the night owing to heavy rains and bad roads."[137]

On Friday, June 11, Day 11 of the contest, the Shawmut reached Colby at 9 a.m., just ahead of Ford No. 2. The Ford stopped only briefly in Colby and left for Goodland, beating the Shawmut out of town to recapture first place. (See Figs. 3.23 and 3.24.) There was little speed involved, however, as the leaders in the New York–Seattle race set about "plowing their way" toward Colorado. "All of Denver['s] automobile-mad citizens will be out to see those cars go through," the *Denver Post* promised.[138] But nobody knew when that would be, the newspaper acknowledged, for a Goodland dispatch stated

> that a veritable lake of mud was spread before the automobiles…. Reports all along the way indicate that weather conditions, which made speed[s] of ten miles an hour look good to the motorists, are being encountered.

Fig. 3.23 The Fords pause before an "Auto Garage & Livery" in Goodland, Kansas.
(HFM P.188.70653)

176

Fig. 3.24 A few onlookers brave the mud to inspect the Shawmut and one of the Ford racers in Goodland. (HFM P.188.70651)

> Even where the rain is not actually falling, it has soaked the roads until the [tire] chains
> are powerless to keep the cars from skidding. The fact that the heroic little cars are going
> ahead in spite of the roads makes the contest doubly interesting to those who want to see
> … a real test of endurance.[139]

Pig Pen to the Rescue

Five inches of rain fell Thursday and Friday, Harper recalled. Friday, the two Fords traveled
15 hours but covered just 90 miles, he added. The Shawmut was the first car through Kanorado,
a border town, at 5:40 p.m.; the Fords followed at 7 p.m. According to a Friday wire report
from Burlington, just 12 miles west of the Kansas-Colorado border:

> The Shawmut car came into town tonight all alone, in the lead of the other entries in the
> ocean-to-ocean race. The Shawmut travelled 90 miles during the day over heavy roads
> and in the rain. She crossed a creek on a railroad bridge, where the Ford No. 1 later got
> stuck in three feet of water. The Shawmut driver will start at daybreak and as the car is
> in excellent condition stands a fine chance of reaching Denver first.[140]

The Fords had several high-water mishaps during the "almost continuous spell of wet weather"
between Kansas City and Denver, Harper recalls in his *Story of the Race*:

> Every day we wore rubber coats and hip boots and pushed through mile after mile of mud. The monotony of this was frequently varied by having to ford a stream where the unusual rain fall had washed away the bridge. Often these swollen streams had beds of quicksand and the car striking them would instantly sink until the body resting on the sand prevented further settling.
>
> Then we thanked our lucky stars that we of the Ford crews were driving light cars. Where a heavy car had to resort to horses and a block and tackle, the two men in each Ford car could pick up their car, place the wheels on planks and proceed across. In all this clinging, clayey mud, quicksand and washouts, neither Ford had to resort to outside power for assistance.[141]

Harper leaves it to his readers to imagine how two men could lift a car weighing 950 pounds. Perhaps the men were traveling with helpers, as in the Missouri mud. Regardless, the Fords on Friday night joined the Shawmut in Burlington, where Denver Ford agent Charles Hendy awaited. Hendy advised the *Denver Post* "that at daylight the cars would try to make the start for Limon." Immediately upon their 7 a.m. departure, the three autos "began to battle with the mire and quicksand," said Denver's *Sunday News-Times*.[142] By all accounts Saturday's 175-mile drive to Denver was among the most difficult of the race.

Shawmut crewmen Earle Chapin, Robert Messer and T. Arthur Pettengill arrived first in Limon—nearly halfway to Denver—"at 12:45 this afternoon, covered with mud," according to a Limon dispatch. "After a short while spent in cleaning the car, the racers left for Byers at 1:25 p.m." Ford No. 2 arrived at 1:33 p.m. and the other Ford two minutes later. Ford No. 2 left at 2:10 p.m.; "the No. 1, being delayed for repairs, left at 3:10."[143] About 35 miles east of Denver, Harper related,

> both Ford cars got into the quicksand in the bed of Sand Creek. We were 30 feet from shore and working in water up to our waists. If we had not had light cars, we might have been there yet, but with the aid of the roof of a deserted pig pen which roof we shoved under the wheels after lifting the back end of the car, we got both cars out.[144]

Mud! Mud! They Say

The Shawmut, however, fell behind with problems of its own, and the Fords eventually separated. As the cars churned on late into the night Saturday, anxious reporters gathered at Denver's Savoy Hotel, hoping to get the latest race results in time for their Sunday editions. But just one car struggled into the city, as the *Sunday News-Times* described it:

> The Ford car, No. 2, in the transcontinental motor car race, arrived at the Savoy hotel at 10:48 last night, several hours ahead of its nearest competitor....
>
> When the Ford car arrived at the Savoy last night it was one mass of mud and its occupants were completely exhausted. The car was taken to the Ford branch garage at 1554

Broadway, after Scott, the driver, had recorded the arrival at the Savoy. At the garage more than two bushel baskets of mud were scraped from the car.

The first car was met at Oakley by Stewart Alkire, the Denver pilot who mapped out the course into Denver. As Driver Scott picked the cakes of dry mud from his hair, ears and neck last night he related the incidents of the worst day's run since the start of the race. He said:

"We were forced to use our block and tackle eight times today to drag the machine from the mud. Throughout the day I used six gallons of lubricating oil to keep the machine in running order, the mud having penetrated to every portion of the engine. I had great confidence in our pilot or I would have spent the night on the prairie rather than brave the mire, mud and water, between us and Denver.

"Unless something happens to our car now there is nothing to prevent our winning the endurance race. This is my first transcontinental trip and had anyone told me that I might live through such a day as this one has been I would not have believed him. I know many of the cars will lose days in the mud east of Denver and some may lose all chance of making a showing."[145]

"Mud! Mud! They Say," cried a headline over the *Denver Post's* account: "The mud-bespattered travelers had touching tales to tell—tales of anxiety and hardship, of hail and rain, of muddy roads—oh! such muddy roads."[146] Mud had gotten the better of the Shawmut and Ford No. 1 crews, who were thus unavailable to tell their tales. (See Fig. 3.25.)

Late Saturday night found the Shawmut "in quicksand fifty miles east of Denver," the Ford No. 1 mired just 20 miles east of the city and the Acme "broken down near Oakley, Kan."[147] The Shawmut and Ford stragglers "spent the night under tarpaulins in their cars, the racers deeply imbedded in the mire, waiting for the daylight hours of Sunday morning that they might pull their cars out with the aid of block and tackle," said the *Rocky Mountain News*.[148]

"A Terrific Hail Storm"

Ford No. 1 finally rolled into Denver at 8:30 a.m. Sunday, June 13, "after having battled with the quicksand near Sand creek and later with the mire and water which were over the hubs of the car for several hours running about twenty miles out of Denver." The belated Ford was "followed closely by the Shawmut," just 30 minutes to the rear.[149] The Acme failed to appear in Denver on Sunday.

"I never saw anything like the condition of the roads which we encountered Saturday between Limon and Denver," Harper told the *Denver Republican*. "I don't see how they could have been worse. In places we had to literally pick up our car and carry it for several hundred feet in order to find ground sufficiently firm to admit of any progress."[150]

Fig. 3.25 Two men on shore prepare to attach a rope from the Buick on high ground to the mired Shawmut in this unidentified scene. Rain and mud slowed all the racers from Missouri through Colorado. (SHS)

All the cars "received a thorough overhauling" before leaving at staggered intervals Sunday for Cheyenne, 113 miles to the north. The first car into Denver Saturday night, Ford No. 2, was the first one out at 12:45 p.m. Sunday. The Shawmut left 15 minutes behind the leader and Kulick and Harper, planning an all-night drive, followed in the Ford No. 1 at 6:30 p.m., the *Denver Republican* said. "Having in mind the hard going ahead," Harper said, Kulick "had decided to go over every part of his car to make sure all was right." Pilot cars accompanied all three racers. Hendy, the local Ford agent, also left Denver Sunday evening, but by train, bound "for various Wyoming points to arrange for pilots for his two cars through his territory."[151]

While the Shawmut and Ford were traveling together just north of Denver, "a terrific hail storm came up which necessitated the men stopping and getting under rubber blankets till it had passed over," according to the *Wyoming Tribune* of Cheyenne. "The cars were together until the muddy stretch between Cheyenne and the Terry ranch was reached where the lighter Ford was able to make better time and thus get to the local checking station first."

The Ford No. 2 reached Cheyenne's Capitol Garage at 7:10 p.m., followed by the Shawmut at 7:40 p.m.[152]

Frank Wright of the Capitol Garage, who piloted the Ford from Denver, "reports that it made a very remarkable run from that city considering the muddy condition of the roads," noted the *Cheyenne State Leader*. "The first hundred miles to Carr [Colorado] was made in four hours flat, and the entire distance covered in six hours."[153] According to a dispatch from Wyoming's state capital:

> There were no mishaps of any kind and after a hurried examination the drivers reported both cars to be in first class condition. The men are all worn out, and both cars will tie up here during the night and while the drivers are resting the cars will be gone over thoroughly and an early start will be made tomorrow. Members of the Cheyenne Auto club piloted the racers into the city and will escort them as far as Laramie tomorrow.[154]

Shawmut Steals March on Fords …

But at 1:40 a.m. Monday, June 14, "while the Ford drivers were peacefully sleeping,"[155] Chapin, Messer and Pettengill crept from their beds, started up the Shawmut and, joined by an Oldsmobile pilot car, headed west for Laramie. "Shawmut Steals March on Others by Early Morning Start," blared a *Denver Republican* headline. The story neglects to mention that the Ford crews had done the same thing in St. Louis. Even if Scott and Smith had wanted to pursue their rivals, the Ford "had to undergo some overhauling" so was unavailable at that hour, said the *Wyoming Tribune*.

"Elmer Lovejoy of Laramie will pilot the Shawmut from that city to Rawlins," the *Wyoming Tribune* added. Lovejoy, who built an auto in Laramie during the winter of 1897–98, knew Wyoming roads as well as anyone. In 1912 he published a route booklet that "indicated every jog, every mud hole to drive around, and even piles of rock to avoid."[156]

The Shawmut reached Laramie at 5:40 a.m., covering the 57 miles through darkness from Cheyenne in four hours, according to a wire report from Laramie. The race leader stopped an hour in Laramie and then left along the popular transcontinental route through southern Wyoming: Medicine Bow, Hanna, Fort Steele, Rawlins and on west.

Its repairs made, Ford No. 2 left Cheyenne at 6:40 a.m. and paused just five minutes in Laramie at 9:40 a.m., "making the run in three hours … and gaining two hours on the Shawmut," the *Wyoming Tribune* reported. The roads between the two cities "were not of the best but far better than was encountered between Denver and Cheyenne." In *The Big Race*, Scott's son writes that Ford No. 2 lost its engine oil when a drain plug popped out at an unspecified location on Monday, June 14. To reach the next town, he writes, Smith and Bert Scott replaced the plug with a piece of wood and the oil with water and melted bacon grease.[157]

A writer for the *Cheyenne State Leader*, inspired by the smaller car's comparatively easy trip to Laramie, suggested that "the Ford No. 2 is playing with the Shawmut," because

> its crew do not appear to be in any hurry and take a full night's rest at every opportunity. It was in Cheyenne 35 minutes ahead of the Shawmut, but left five hours later, after the crew had slept for nine hours at the Inter-Ocean. The Shawmut crew upon reaching Cheyenne had been without sleep for three nights, and only slept for about three hours in this city.
>
> The little Ford gained two hours on the Shawmut between this city and Laramie, the trip over the hill being covered in four hours [sic], while the Shawmut made it in six [sic]. Running over soft, rough roads seems to be more difficult for the larger car, and the little Fords are running ahead at most any time they please, and the crews are taking plenty of time to rest and for the cars to be overhauled.[158]

Not as red-eyed as the writer assumed, the men in the Shawmut actually slept at a farmhouse south of Colby Thursday night, at Burlington Friday night and under a tarp east of Denver Saturday night.

Ford No. 1, Acme Play Catch-Up

Ford No. 1, which left Denver early Sunday evening, had pushed through the mud until 11:30 p.m., when it stopped for the night at Greeley, Colorado, about 50 miles south of Cheyenne, dispatches said. "They found the going extremely hard with the muddy roads and the darkness to overcome, but did well considering the conditions encountered."[159]

Kulick and Harper left Greeley at 6 a.m. Monday and, "after driving through a hard hail storm and roads that were little more than rivers of mud,"[160] reached Cheyenne shortly before 9 a.m. They left at 11:15 a.m. and reached Laramie at 2:10 p.m. Upon leaving Cheyenne, Harper recalls,

> we noticed No. 3 the China-Japan fast mail on the Union Pacific just pulling out.... The schedule for this train calls for arrival at Laramie at 2:15 p.m. The train winds in and out around the mountains—the automobile road goes straight over them. We left with the train and it pulled away. Five times we met and the passengers displayed a continually increasing interest. They began to watch for us. We arrived in Laramie five minutes ahead of the train.[161]

As Ford No. 1 left Cheyenne late Monday morning, the Acme rolled into Denver, "after being delayed at Quinter, Kan., by the breaking of its Jack shaft,"[162] caused by hitting a tree near Centralia, Missouri. As in most other chain-driven cars, the Acme copied the bicycle's system of power transmission. That is, the car's engine turned a transversely mounted jack shaft, having at each end a sprocket similar to a bicycle's drive sprocket. A chain connected the sprocket on each side of the Acme to the hub of the rear wheel immediately behind it. After "a hasty overhauling"[163] in Denver, the Acme would pursue the leaders. As the *Denver Times* reported the Acme's arrival:

> The machine carried about a half a ton of mud caked on the wheels, engines [cylinders] and running gear. F.H. Hempstead [J.A. Hemstreet], the driver, and his three companions were begrimed and spattered with mud and showed the hardships of the trip.

> "We've lost a lot of time, but we're going to win anyhow," he said on his arrival at the Savoy hotel. The car was cleaned here and will depart for Cheyenne this afternoon.[164]

... Fords Steele Back Lead

At about 5:30 p.m. Monday, the Ford No. 2 passed the Shawmut near Hanna, 80 miles northwest of Laramie, various reports indicate. But the Shawmut was leading when it reached the North Platte River at Fort Steele, 25 miles farther west, at 6:30 p.m., according to *Motor World*. It was to protect the builders of the Union Pacific Railroad that the U.S. Army in 1868 established Fort Fred Steele, as it was formally known. Abandoned in 1886, the fort by 1909 had largely fallen into ruins.[165] So had the wagon bridge across the river at Fort Steele, as the 1908 New York–Paris racers learned. Accordingly, the Thomas racer had crossed the river on the ice, an option now unavailable. The New York–Seattle drivers had another option for crossing the deep river: the Union Pacific railroad bridge. But when the Shawmut prepared to cross the bridge, *Motor World* related,

> a vigilant station agent held it up by force, demanding that a permit be shown. A long delay ensued before the general manager at Omaha could be communicated with, the two Ford cars arriving meanwhile and starting gayly over without interference, since their drivers thoughtfully had provided themselves with the necessary credentials and permits.[166]

More likely, the Ford agent in Cheyenne or Rawlins requested the permits. Regardless, while the Shawmut crew waited helplessly on the east side of the river, Ford No. 2 crossed the bridge unopposed and drove on to Rawlins, where it stopped for the night at 8:55 p.m. As a dispatch from Rawlins revealed Monday night, Scott and Smith had little cause to celebrate their lead, for travel prospects were grim: "The red desert country is almost impassable owing to heavy rains. Bridges and culverts are in bad shape from here west." Harper later observed: "When it rains in Wyoming, it does not make much mud. It just pours down the mountain sides and cuts sluiceways across the road at intervals of about 50 feet. These ditches run about 18 inches deep and offer considerable difficulty to automobiles."[167]

Itala "Considerably Damaged"

Late Monday found the fourth-place Acme crew "stuck in the mud" at Pierce, Colorado, about halfway between Denver and Cheyenne.[168] The Itala racers, perhaps taking risks in their haste to catch up, had an accident Monday in central Missouri, where the leaders crossed the Missouri River six days previously. As reported in a Monday dispatch from Marshall, Missouri, near the scene of the accident:

> The Itala car, fifth in the ocean-to-ocean automobile race, arrived today, considerably damaged. While crossing the Missouri River on the railroad bridge at Glasgow the car was struck by a freight train, which knocked off the gasoline tank and broke the front axle. The car likely will reach Kansas City tomorrow.[169]

To continue after the accident, one account said, "The Itala crew extended a piece of rubber tubing from a portable gasoline can to the carburetor, and from then on into Kansas City, where the tank was repaired, there was always a front seat passenger with a can of gasoline in his lap, nursing the fuel through a tube into the engine."[170]

Fords Break Wheels

The Shawmut crewmen spent Monday night at the North Platte River. They were still awaiting permission to cross when Kulick and Harper—who had spent Monday night at Medicine Bow, 45 miles east—pulled up at 9 a.m. Tuesday, June 15. They, too, bumped across the railroad ties as the Shawmut crew looked on despairingly. Finally, at 10 a.m., UP headquarters in Omaha, Nebraska, wired permission for the Shawmut to use the bridge. Race organizers, of course, should have arranged it so all cars could cross this bridge speedily; the Shawmut factory called the 15½-hour delay at Fort Steele unfair. Accordingly, it would file a formal protest.

As it turns out, the Fords' advantage was illusory. For on its oversize 40-inch wheels, the big Shawmut crossed the trestle unscathed. The lighter Fords had smaller wheels, as Harper described in *Story of the Race*:

> To get into Rawlins necessitated using the railroad ties for a mile, this including the approach to and the railroad bridge over the [North] Platte River at Fort Steele. The track was not ballasted and the ties 15 to 18 inches apart. Our 30 inch wheels hit every separate tie a distinct and separate bump and each car came into Rawlins with a broken wheel. It delayed us twelve hours making repairs and we were now 2nd and 3rd.[171]

The leading Ford No. 2, which made its getaway from Rawlins at 6 a.m. Tuesday, "broke down six miles out, and was towed back for repairs," reports said. The car slid into a ditch and broke a front axle, which was welded in Rawlins, according to the Ford company. Ford No. 1 reached Rawlins at 10 a.m., an hour after crossing the river, "badly crippled, having broken a wheel while crossing the Fort Steele Bridge."[172] The two Fords thus spent much of Tuesday in the shop at Rawlins, leaving together at 4:45 p.m. The Shawmut, meanwhile, reached Rawlins at 11 a.m. and headed west again at noon. By reaching Wamsutter at 4:25 p.m., the Shawmut "gained a lead of forty miles before the Fords got away from Rawlins." The three leading cars drove on into the night.

Stalled in fourth place, the Acme wasn't going much of anyplace Tuesday night. A Tuesday wire story from Cheyenne said the Acme "was stuck in the mud at Pierce, Col., 40 miles south of here, all day, and then, after moving 13 miles, got stuck again." A more foreboding account in *Motor Age* said the Acme broke an axle at Pierce "and will be hung up until a new part reaches there from Denver."[173] The auto journal was vague about exactly what broke; replacing the entire axle would have been against the rules. A post-race article in the *Seattle Post-Intelligencer* asserted that at Pierce the Acme broke its jack shaft for the second time.

Together at Opal

Ahead of the leaders—now nearly halfway across Wyoming—lay Rock Springs, Green River, Granger, Opal, Kemmerer, Cokeville and the Idaho border. Their route across southern Idaho would take them through Montpelier, Pocatello, Twin Falls, Boise and smaller settlements. On Wednesday, June 16, as the weather and roads dried up, both Fords beat the Shawmut into Rock Springs. But the big car left in first place at 12:25 p.m. while the runabouts stopped again to repair wheels damaged on the Fort Steele railroad bridge, the *Montpelier Examiner* reported.[174]

They left Rock Springs about three hours behind the Shawmut, which, unaccountably, drove the 45-mile stretch to Granger at a snail-like pace, arriving at 5 p.m. Ford No. 2 stopped in Granger for 12 minutes at 6 p.m. and zoomed on west in first place while the Shawmut dallied until 6:25 p.m. Ford No. 2 stopped for 20 minutes and left at 7:55 p.m., 90 minutes behind the Shawmut and in third place.

"All these cars expect to make Opal to-night, twenty-five miles," according to a dispatch from Granger. "The roads are dry and dusty, and the cars are apparently in good condition."[175] At Opal, Harper related, "all three cars met and[,] a storm starting, all crews slept there until daybreak." Early Thursday, June 17, the Ford drivers took the lead for 15 miles into the next town. "At Kemmerer just as we finished breakfast, the Shawmut crew came into the lunch room. These but serve to show how close and exciting the race became from time to time." Hoping to make the race closer and more exciting, the Itala crew left Kansas City early Thursday, passed Salina after 7 p.m. and headed on into central Kansas. "Rains have made the roads heavy and progress is slow," a dispatch advised.[176]

The tightly bunched leaders headed north to Cokeville, near the Idaho border, where Ford No. 2 broke down. According to the *Pocatello Tribune*, Scott and Smith spent the day at Cokeville repairing the unspecified damage. Harper's *Story of the Race* refers only to "an accident at Cokeville." Years later, however, the Ford Motor Company revealed that "Ford No. 2 reached Cokeville with a broken connecting rod and went into a Ford dealer garage for repairs."[177] The Shawmut fell behind because it, too, had a breakdown Thursday near Cokeville, according to one report, which said the Shawmut crew quickly made repairs and hurried on.

Ford No. 1 Reaches Pocatello

Thus Ford No. 1 reached Montpelier alone and in first place at 4:04 p.m. Thursday. Alerted to the leader's approach, an enterprising local motorist jumped into his Ford to meet Harper and Kulick east of the city. He "made a great reputation as a speedy driver, keeping close behind the racing machine coming into town," the *Montpelier Examiner* reported. "A new front wheel and springs were put on the car and at 6:30 p.m. it left for Pocatello, expecting to make the 100-mile run by midnight," related a Montpelier dispatch in the *New York Times*.[178] The *Montpelier Examiner* gave the departure time as 5:50 p.m. The second-place Shawmut reached Montpelier at 7:05 p.m. and set out at 8:20 p.m. for the run through Bancroft and Soda Springs to Pocatello, the *Examiner* said.

Kulick and Harper reached Pocatello well past midnight, reported the *Pocatello Tribune* of Friday afternoon, June 18:

> Ford car No. 1, the leading automobile in the New York-to-Seattle race, arrived in Pocatello this morning at 2 o'clock, and after undergoing a general overhauling and clean up, departed for the west at 9 a.m....
>
> It is a very light looking machine to make such a hard trip, but is standing up under the strain remarkably well. It is not a very attractive looking car, with its coating of mud and dust, and makes as much noise as a threshing machine owing to lack of a muffler....
>
> The car was piloted from Soda Springs to Pocatello by Chris. Woodall, who knows every inch of the road. He enjoyed the night trip last night immensely and said that in coming down the Portneuf canyon, a speed of close to 20 miles an hour was maintained. "We were not troubled by those ditches and washouts, at all," he said, "as we simply jumped over the low places and hit only the high spots."[179]

Woodall was alone in his enthusiasm. As Harper recounted the adventure:

> If you want excitement, try that run into Pocatello from Opal.... after a heavy rain. If you do not get enough in the day time, try it after dark. But look out for Nugget and Bancroft Canyons. There are down grades there, several of them up to 1,000 feet in length, averaging 30% where the roadway in places is six inches wider than the car tread and a slip means dashing hundreds of feet onto the rocks below.[180]

Shawmut, Acme Lose Bearings

The Shawmut crew had some excitement of its own, breaking down Thursday night about four miles east of Bancroft, some 55 miles from Pocatello. "A bearing on the front axle gave way, and it was necessary to bring the broken part to the Short L[i]ne shops in this city for replacement by casting," Friday's *Pocatello Tribune* said. "The work was hurried as much as possible and the broken part was taken down to Bancroft on No. 8 [a train] this afternoon."[181] Another report said the broken part was a bearing race, not the bearing itself.

Ford No. 2, in third place, also made a nocturnal drive. From Cokeville, where it had broken down, Scott and Smith resumed their journey Thursday evening. The car reached Montpelier at 7 a.m. Friday, passed the broken-down Shawmut and, at about 2 p.m., reached Pocatello, where it "spent an hour in cleaning up and proceeded on its way rejoicing."[182]

The Acme crew—which left Denver for Cheyenne Monday but had been stuck for 3½ days in northern Colorado with a broken jack shaft—spent Friday rejoicing, as well. "Acme Auto Finds Cheyenne," cried a headline in Friday's *Wyoming Tribune*, which said the car arrived at 6 a.m.,

> stopping at Dinneen's garage for gasoline and examination. At 6:20 the car was on the way again for Laramie with Lyle Branson of the Dinneen garage as pilot. The Acme lost a great deal of time last night by getting lost. It had no pilot between Carr and Cheyenne and the driver lost his bearings between Pierce and here. It lost several hours by this mishap.[183]

Fool Fouls Ford

No such debacle had befallen the two 20-horsepower runabouts. That's because, as the Ford company arranged it, a dealer or motorist familiar with local roads piloted each Ford racer a certain distance and then passed on the responsibility to another local expert. Freed from navigating, the drivers could concentrate on winning the race. But in Idaho, the seemingly foolproof system encountered—well, a fool, Harper would argue. At Pocatello, recalls Harper,

> We began to plan on what to do with our money. We might better have saved our breath. For then did we pick up … a pilot … whose previous experience must have been largely confined to piloting schooners over the bar. He was sure bone-headed and he certainly proceeded to lose Ford car No. 1.
>
> He got us into the north end of the Great American Desert where we averaged four miles an hour. Then when we overcame this, he lost the road again and [Friday] night instead of hustling toward Boise, we slept in the sage b[r]ush out on the plain, *sans* gasoline and oil and 55 miles off the roads.
>
> By the time we had walked three miles, pumped a hand car six more, flagged a passenger [train] and proceeded to Shoshone, Idaho, for fuel and oil, then borrowed a car to carry it back and again started for Twin Falls, we had lost twenty-four hours.[184]

Harper and Kulick had faithfully followed their guide's directions out of Pocatello Friday, traveling along the north side of the Snake River, and wound up spending the night "broken down in the sands near Minidoka," according to the *Pocatello Tribune*.[185] At Pocatello they were six hours ahead of Ford No. 2, which followed on the south side of the river—evidently the better route of the two. The delayed Ford No. 1's arrival on Saturday, June 19, caused a stir in Shoshone (See Fig. 3.26), according to the *Shoshone Journal*:

> Ford car No. 1 which was in the lead in the ocean to ocean race when it reached Pocatello, took the road to Shoshone instead of Twin Falls and by so doing lost valuable time. It ran out of gasoline east of this city and the drivers had to come here for a new supply. When the car rolled into town after the noon hour it was immediately surrounded by a crowd. Photographer Merrifield was on hand with his big camera and within a few hours ... he had perfected a splendid photo of the machine and crowd.[186]

Fig. 3.26 With Kulick driving and Harper alongside, the Ford No. 1, right, pulls into Shoshone, Idaho, far off the scheduled route. (July 15, 1909, Ford Times/HFM 0.3891)

From Shoshone, Ford No. 1 had to backtrack south through Jerome, Idaho, to check in at Twin Falls on Saturday, thus losing more time, according to the *Idaho Daily Statesman* of Boise. According to a 1959 Ford Motor Company account, "Frank Kulick, only one of the four Ford drivers now living, still swears the guide got them lost on purpose."[187]

Ford No. 1 actually picked up the confused pilot at American Falls, about 25 miles west of Pocatello, Harper said while recounting the tale in *Ford Times*: "Good pilots had been secured and instructed to be in readiness, but the Ford No. 1 was now ahead of any schedule. It arrived in American Falls two hours ahead of the pilot. Rather than wait, for time was precious, another pilot equally well recommended, was secured and the journey continued."[188]

Patience was a virtue for Ford No. 2, as "our car never did get lost," Smith said in a 1951 interview. "We waited for our pilot all the time, but when we did have to wait an hour or so, well, we'd wait rather than get lost."[189]

Sleepless Till Seattle

Meantime, the Shawmut crew in Bancroft installed the new casting it received by train, and roared into Pocatello at 7:20 p.m. Friday. The car headed west at 9:40 p.m., "arriving at American Falls at 10:55," reported the *Pocatello Tribune*. "J.B. Trist and Theo. Turner piloted the car through to the latter place. During the trip, an average of better than twenty miles per hour was kept up."[190] The Ford No. 2 passed the wayward Ford No. 1 sometime during its nighttime drive to assume first place.

The Shawmut left Twin Falls at 3 p.m. Saturday, evidently ahead of Ford No. 1, as well, and thus in second place. The Shawmut crew would drive all night to reach Boise, one account said. The Shawmut reached Buhl, about 20 miles west of Twin Falls, at 4:15 p.m. "and left in a pouring rain at 5:20 o'clock," according to a dispatch from Buhl. "It was piloted to Clear Lake ferry by Seymour Fairchild. Ford No. 1 is expected here at 1 o'clock Sunday morning.... The roads are bad on account of the rain and going is slow."[191]

M. Robert Guggenheim, organizer of and referee for the cross-country race, found the going decidedly fast on Seattle's streets Friday. Guggenheim, who had pledged that the racers would obey all speed limits between New York and St. Louis, "was arrested by the police here on the charge of exceeding the speed limit with his racing car," according to a dispatch from Seattle. "He was taken to police headquarters today, where he deposited $50 bond for appearance in court."[192]

From central Idaho, the official racers would find Bliss—the town, that is—Mountain Home and Boise. J.D. Moore, manager of Boise's Intermountain Auto Company, met Ford No. 2 early Saturday afternoon at Bliss, 95 miles from Boise. With Moore acting as pilot, the new leader in the New York-Seattle race reached Boise over wet roads at 8:05 p.m. Saturday (See Fig. 3.27), according to Sunday's *Idaho Daily Statesman*:

> The little car was completely covered with mud on its arrival here. Mr. Scott, who is in charge of the car, said last evening that he expected to leave for the west at 3:30 o'clock this morning. "We have had little sleep and are not looking for sleep," declared the driver. "We will sleep when we get to Seattle and plenty, too," he added.
>
> Upon registering here the two men were presented with five $20 gold pieces, the gift of the Boise Automobile club to the drivers of the first car to reach here in the race....
>
> Scott and Smith were a tired pair when they turned in at the Idanha [Hotel] for a few hours sleep last night. They bathed soon after arrival, ate supper, and went to bed.[193]

Fig. 3.27 Nattily attired residents of Boise pack the Intermountain Auto Company's garage to greet Smith, foreground, and Scott, whose Ford No. 2 arrived in first place on Saturday evening, June 19, 1909. (July 15, 1909, Ford Times*/HFM)*

Late in the race, as Smith recalled years later, "We kept right on going all the time; one would sleep and the other would drive."[194]

Ford No. 2 Lengthens Lead

Henry Ford "said at the start that he had entered his two cars to win first and second," reported the *Seattle Post-Intelligencer*. Accordingly, Ford wired R.P. Rice, Seattle branch manager, "to see that the cars finished in that order." Rice thus dispatched Harry Disher to leave Seattle's Ford garage for the east early on Sunday. Disher's orders were to cross the Cascade Mountains through Snoqualmie Pass, 50 miles east of Seattle, "and continue until he meets the first Ford car on the road, when he will turn and pilot the car back to the finish."[195]

Three hours later than planned, Moore led the race leaders out of Boise at 6:30 a.m. Sunday, June 20, bound for Payette and Weiser, Idaho, near where the car would cross the Snake River into Oregon. "Ford No. 2 is off to a good start," commented the *Idaho Daily Statesman*. "It is a day's travel ahead of the other contestants and Scott and Smith, the gritty drivers, declare they will be the first ones to reach Seattle."[196]

Ford No. 1 arrived in Boise eight hours behind the leading Ford but ahead of the Shawmut at 2:30 p.m. Sunday. "The car was given some slight repairs at the Intermountain garage where all the cars check," and left about 7:30 p.m.[197] Unaccountably, the Shawmut consumed 24 hours in traveling the 100 miles from Twin Falls to Mountain Home, where it arrived at 3:30 p.m. Sunday. It traveled the 50 miles to Boise, which it left in third place at 10:15 p.m., bound for Baker City, Oregon.

In the meantime, the leading Ford had driven 160 miles over bad roads to Baker City on Sunday, the *New York Times* said. "Although for the last two days heavy rains have fallen over Eastern Oregon, rendering the roads about impassable in places, the Ford car No. 2 in the New York to Seattle automobile race arrived in Baker City" at 3:45 p.m. "Both men were in good condition and the machine was in perfect running order," reported Monday's *Baker City Herald*. "After taking on a supply of gasoline and lubricating oil the car left for Walla Walla," a 130-mile trek.[198] Most of the racers traveled through Baker City (now called simply Baker) and Pendleton in northeastern Oregon to reach Walla Walla, Washington.

Ford Agent Predicts 1-2 Finish

Charles Hendy, the Denver Ford agent who was traveling in advance of the racers by train, greeted the Ford No. 2 crewmen in Baker City. "There is little doubt that No. 2 will arrive in Seattle first," he told the *Baker City Herald*. "Ford No. 1 lost her way … in Idaho and traveled 100 miles that would have been unnecessary. The Shawmut is coming along and will finish third in my opinion…. We are trying to demonstrate that the light car is the one for all kinds of travel and I believe we will do it."[199]

But the Shawmut, battling hard for second place, reached Baker City at 8:50 a.m. Monday, June 21, some 17 hours behind the leader, "and after registering, took off toward Walla Walla," the *Herald* reported. "Ford No. 1 followed the Shawmut about 20 minutes later." Earle Chapin dropped out of the race at Baker City, "owing to illness," said a report, without elaborating.[200] This left Messer and Pettengill to drive the Shawmut to Seattle.

For the third time in as many states, the trailing Acme on Friday broke its jack shaft, this time at Bosler, Wyoming, 20 miles north of Laramie. The crew repaired the damage and traveled 100 miles to reach Rawlins on Monday, June 21. After a 50-minute stop, the car—"now in excellent condition"[201]—continued west. Driver George Salzman "is gamely sticking to his work, and declares he will finish with his car," said a wire report from Wyoming.[202] The Acme left Rawlins for the west at 2 p.m. Monday, with three states between it and the leader.

No. 1 Slips Out, Loses Way

Ford No. 2 arrived first in Walla Walla at 6:45 a.m. Monday and left at 8:15 a.m., hoping to cover the 291 miles to Seattle by Wednesday morning. Ford No. 1 arrived in Walla Walla at

6:10 p.m. and the Shawmut at 7:05 p.m. "The drivers and machinists with the cars were all in good condition, weary from the long journey," according to a report from Walla Walla. "All the cars were oiled up and gone over thoroughly here, as the drivers intend to push the cars right through."[203]

The two trailing autos sought repairs at the same garage, the dispatch said. When the Shawmut pulled out of the garage at 9:10 p.m., "it was only to find that Ford No. 1 had slipped out of a back door 10 minutes before. Fifteen miles out the Ford met with an accident and had to send to Walla Walla for repairs. The Shawmut forged far ahead in the interim."[204]

Reporting on its progress as of Monday, the *New York Times* praised the Shawmut auto: "The performance of the Shawmut car has been remarkable, for throughout the long and trying journey the car has labored under the handicap of being without adequate equipment. Any sort of serious accident would put the Shawmut out of competition, for it has not had the support of a factory such as the Ford, Acme, and Itala cars have had. When the Shawmut left New York automobile experts prophesied that it would be the first car to drop out of the contest."[205]

From Walla Walla, most racers followed the route to Seattle through Wallula, Prosser, North Yakima (now simply Yakima), Ellensburg, Easton, Snoqualmie Pass and North Bend. With a pilot, Ford No. 1 crossed the Columbia River at Wallula, intending to travel to North Yakima through the Horse Heaven Hills. That was the plan, anyway, Harper would recall:

> At Wallula we picked up another bone-headed specimen for a pilot. This road juggler lost us in what is known as the Horse Heaven Country, and when we should have been in Prosser, Washington, we were in Mottinger. Any jury in the land would have brought in a verdict of justifiable homicide if we had followed our inclination in regard to the excess baggage that had hired out as a pilot to us.
>
> We had to travel the ties on the S.P. and S.R.R. [Spokane, Portland & Seattle Railway] for eight miles to Plymouth. Part of the distance was through a tunnel dark as night and just wide enough for that single track. The road was entirely lost in the sand. In that country the sand drifts and blows with every puff of wind—a sand storm that was a veritable blizzard of sand had raged around us for hours. To have stopped anywhere on that desert for the night would have meant being buried in the sand. In places it had blown in piles as high as a three story building.[206]

"A Mass of Flames"

Leading the pack, Ford No. 2 had reached Prosser at 7:30 p.m. Monday. "After filling gasoline tanks and eating supper Scott and Smith pulled out for North Yakima." They left at 8:45 p.m., accompanied by a pilot in another Model T, according to a dispatch from Prosser.

Just as they were ready to start, some bystander carelessly struck a match on the back of the car to light a cigar, and in an instant the rear of the car was a mass of flames, and the crowd of 200 scattered in lively fashion. The hand chemical of the Prosser fire department was rushed out and quickly extinguished the fire before any harm was done.[207]

Other accounts say the fire started when someone struck a match on the gas tank while the car was refueling. Naming names, one report claimed the fire was caused "by the too willing assistance of Editor Watson of the *Prosser Bulletin* in filling the gasoline tank of Ford car No. 2." Reports of damage ranged from slight to serious. Though not a witness, Harper contended "fifteen gallons of gasoline and the major portion of the outfit were burned and the tank sprung a leak, causing considerable inconvenience for the balance of the journey." *Automobile* agreed that "the tank's usefulness was ended … and for the rest of the journey the fuel had to be carried in a can on the footboard."[208]

When Ford No. 2 reached North Yakima at 2:30 a.m. Tuesday, June 22, "Mechanic Charles Smith was asleep in the car," according to a dispatch. "After remaining two hours for minor repairs they resumed their journey to Ellensburg. The car is in fairly good condition. Bert Scott, the driver, and Smith are nearly exhausted by the long run."[209] The leading car reached Ellensburg at 7:45 a.m. and Easton—about 75 miles from Seattle and the gateway to the Cascade Mountains—at 1 p.m. Tuesday. They headed west from Easton.

Henry Ford and son Edsel arrived at Seattle's Rainier-Grand Hotel Tuesday night, reported Wednesday's *Seattle Post-Intelligencer*. "Without bothering to go to his rooms he told the clerk he would not occupy them last night as he did not expect to get any sleep. Immediately after registering Mr. Ford and his son started in a speedy little Ford runabout driven by Manager [R.P.] Rice, of the local Ford agency, to meet the incoming Ford racer and accompany it into Seattle."[210]

Itala Quits Race

It appeared to Guggenheim, the Itala owner and race referee, that the Shawmut and two Fords would all finish; only the order was still in doubt. He thus made two decisions Tuesday. One concerned the Itala, crippled since colliding with a freight train in Missouri. The car reached Denver on Monday, eight days behind the leaders, and continued north to Wyoming. Guggenheim, however, decided to withdraw the car at Cheyenne. His second decision Tuesday was to call on Seattle's mayor "for police assistance in affording a clear track for the racers, and the mayor has instructed Chief Ward to detail as many men as possible to see that the flying machines are not interfered with."[211]

The Shawmut arrived in Ellensburg at 5:17 p.m. Tuesday, eight or nine hours behind Ford No. 2, according to the "last authentic report" received in Seattle that day. The big car "tore into Ellensburg, stopped ten minutes for gasoline, and tore out again, the driver, T.A. Pettengill, and mechanic, R.H. Messer, fighting desperately to overcome the Ford's lead."[212] In Ellensburg, according to a dispatch, "the driver said he stood a good chance to catch the Ford car in the mountains."[213] Ford No. 1 was in third place, its whereabouts unknown, Tuesday night, according to the *Seattle Post-Intelligencer*.

Tuesday night, Ford No. 2 was climbing the Cascade Mountains "and is out of reach of telegraph or telephone." The leaders planned to cross Lake Keechelus on a ferry and then run all night. "The roads over the Cascade Mountains are bad, snow being from one to four feet deep. Men worked all night clearing a path for the racers," the *New York Times* said.[214] At 3,004 feet, Snoqualmie Pass was low compared to the Rocky Mountains of Colorado and Wyoming. Still, it was often clogged with snow well into the summer months. As Scott avowed to Seattle reporters at the finish line on Wednesday:

> The Snoqualmie pass was the worst piece of going that we encountered on the road. We crossed Lake Keechelus at 5 o'clock last night, and as we toiled up to the summit we were soon floundering in snow which, in places, was five feet deep. We reached the summit at 8 o'clock and continued on our course until 9:30, when we stopped for the night at a railroad camp, and slept in a siding car until 2:30 this morning, when we started off again. In many places we had to dig our way out of the snow, and practically climb over logs which lay across the road.[215]

"At that time of the year there was ice on top of the snow," Smith, Scott's companion in Ford No. 2, recalled years later. "We rode along this pass until we got pretty nearly over it; then we broke through the snow. We could hear fellows pounding in the distance. They were on a railroad across there. We got the section gang, and they helped shovel us out. It wasn't very far that we had to shovel." The drivers also averted a delay at Lake Keechelus, according to Smith: "The fellow on the boat, who was going to ferry us across, was going to wait until the rest of the cars came. We had to get a sheriff to make him take us across; otherwise, we would have waited till they all got there."[216]

At 9:55 a.m. Wednesday, Ford No. 2 passed through North Bend, from which "the remainder of the distance is down hill over good roads," according to a report. But where was the Shawmut? At 10:30 a.m., when Ford No. 2 reached Issaquah, just outside Seattle, it was a mere 12 minutes ahead of the Shawmut, "which has made a remarkable gain on its rival," according to another dispatch. Remarkable indeed! For, as another account revealed, "it is supposed that the car following the Ford belongs to some private party, and was not the racer."[217] Scott and Smith, in fact, rolled into Seattle with a comfortable lead.

"Masses" Greet Winning Ford

The celebrating began as the leaders reached the smaller towns near Seattle. At Renton, a few miles southeast, "a lane of people ... crowded the business streets," a report said. "The little racer, dusty and battered[,] looked anything but impressive in the company of the big, bright touring machines of the escort."[218] A procession of 150 automobiles joined the winner for the final march into Seattle, according to one estimate. Around its radiator, Ford No. 2 carried "a great wreath of blossoms placed on it at North Bend."[219]

"I've never seen such a crowd in my life," Smith said in a 1951 interview. "Outside Seattle, the farmers were all out.... We saw that a farmer had hung out a wreath made out of clover, hay, and other things off the farm. He had hung it out on a pole. I put my arm through it, and we hung it over the radiator."[220]

Scott was driving when he and Smith reached Seattle. In the city, First Assistant Fire Chief W.H. Clark, driving the fire department's "patrol automobile," jumped to the head of the procession. Next, as the *Seattle Post-Intelligencer* described it, came

> M. Robert Guggenheim in his racing machine, with Mrs. Guggenheim and [L.]A. Walker, secretary of the contest committee. Following the donor of the trophy came "Scotty" and "Smithy" in the winning car, while in their wake was R.P. Rice, manager of the local Ford branch, with [Henry] Ford, president of the Ford Motor Company, and his son.

> The continuous blast of the pneumatic siren from the fire patrol and the loud "barking" of the [Guggenheim] Allen-Kingston racer caused the crowd to part and allow the cars a clear right-of-way. The street crossings were guarded by patrolmen, and the cars hit such a hot pace through the city that many of the following machines dropped out of the procession.

> Those who held points of vantage on roof-tops and on telegraph poles notified the gathered crowds below of the approach of the cars, and the masses retreated to the sidewalks.

> When Mr. Guggenheim, referee of the race, reached the gate, he jumped from his car and ran to the gate where he took the time of the finish when the winner broke the tape. [See Fig. 3.28.] 10,000 people had gathered at the fair grounds to see the finish of the big race.... Entering the grounds the winning car was driven to the Stadium, where it completed five exhibition laps....

> A place of honor was assigned to the winning car in the Mines building, where the Ford was driven from the Stadium. Scott drove the car up the flight of stairs to the entrance of the building, but on reaching the landing it was discovered that the door would not allow the entrance of the machine, so he backed down and entered the building through another doorway. [See Figs. 3.29 and 3.30.]

> Scott and Smith smilingly granted poses for an army of amateur photographers, but politely overlooked the notebooks and pencils that were handed them for signatures.

> The party was taken to the office of J.E. Chilberg, president of the exposition, by Richard Eskridge, chairman of the entertainment committee, and an informal reception was tendered the victors before they had an opportunity to remove their war paint....

> Scott and Smith left the exposition grounds with Mr. Ford and went to the Rainier-Grand, where they were turned over to the tender mercies of barbers, tailors and other artisans of

Fig. 3.28 As two men hold a tape across their path, Smith, left, and Scott enter the grounds of Seattle's Alaska-Yukon-Pacific Exposition on June 23, 1909, in their Ford No. 2 racer. The clock suspended above the exit turnstiles records the time as 12:53 p.m. (UWSC Nowell X2203)

a kindred sort until the "lightning change" brought them into closer semblance to the ordinary mortal. Invitations to balls, theater parties, receptions and banquets were gracefully declined and the tired men sought the quiet of their rooms to recuperate before the reception committee begins its official campaign.[221]

"Stimulating Sentiment for Good Roads"

The *New York Times* sang the praises of the lightweight Ford:

One thing—and a very important one—that was not taken into consideration by those who prophesied defeat for the Fords, was the fact that the little cars had more horse power to comparative weight than the larger and racier looking machines. The little cars could push through the tangled vegetation of the plains more easily than the heavier cars, and when it came to traveling over rough roads the small cars again had the advantage of lightness and power.

Had it been merely a question of speed over smooth and unobstructed roads the larger cars would have triumphed. The Acme, the Itala, and the Shawmut would have left the Fords far in the rear. It was the mud holes and the stiff sage brush of the plains that defeated the heavy cars.[222]

196

Fig. 3.29 From left to right, Smith, Scott and Guggenheim pose with the Ford racer in the Mines Building. (UWSC Nowell X2192)

Fig. 3.30 Another view of the soiled Ford Model T race car shows its ill-fitting hood, listing body and tattered tires. A bucket under the car catches oil drips. (UWSC Nowell X2194)

Over a mere three weeks, many race observers reversed their opinion of the Model T, the automaker noted: "In New York they [the cars] were sneered at, in Buffalo it was to laugh, in Denver to wonder, Pocatello to fear and in Seattle to marvel, for the cars stood this hardest test better than cars which sell for from five to ten times the Model T price."[223]

Though most accounts placed the finish time at 12:55 p.m.—precise to the second, the Ford Motor Company said 12:55.35—a photograph of Ford No. 2 entering the exposition casts doubts on that time: a large clock in the photo clearly reads 12:53 p.m. The official finishing time was two minutes later, regardless, meaning Ford No. 2 had traveled the 4,106 miles in an elapsed time of 22 days, 55 minutes.[224] It thus averaged 186 miles per day and 7.76 mph over its elapsed time. Ford No. 2 replaced a rear tire in Chicago and another one in Walla Walla, press accounts reveal. "Had it not been for the fact that we were obliged to use chains on the rear wheels, we would have arrived here with New York air in all four of the tires," Scott said.[225]

"A lot of people wanted to know if we had tire trouble," recalled Smith, who became an instant celebrity because of the race. (See Fig. 3.31.) "It was a light car; we had no tire trouble." The 1909 Model T did have a weak link, however, he revealed decades later:

> Our axle was one of the things we had to change bearings in, because we started out with babbitt bearings in the rear axle—not even roller bearings but babbitt bearings. Pounding on the road would elongate them and, just as soon as they would get elongated too far, we'd pull up to a tree, pull them off and stick new ones in. We carried spare parts because we knew the axle wouldn't stand up; that is, the bearings wouldn't stand up because of the pounding all the time.

Fig. 3.31 An admirer pasted a photo of Jimmy Smith on a envelope and mailed it. "This letter was delivered O.K.," according to Ford Times. *(Dec. 1, 1909,* Ford Times/*HFM)*

That feature was changed on the car the next year. Then they all had roller bearings and holding action. That was just a babbitt bearing in there, just the same as you'd have in a lineshaft that old machine shops used to use.[226]

Scott and Smith in Ford No. 2 failed to break the transcontinental record of 15 days, 2 hours, 12 minutes, set in 1906 by a Franklin auto. But they did trim 2 days, 17 hours, 50 minutes from the east-to-west speed record that Christian D. Hagerty and Richard H. Little set in a Buick in 1906. "The winners have accomplished a notable and meritorious feat," allowed *Motor World*, which praised the racers' performance despite opposing the race.[227]

"I do not think anything could have happened to stimulate a greater sentiment for good roads than the transcontinental contest just finished," Scott told reporters. "Where roads were bad the sentiment of regret prevailed, and the desire and promise was for improvement. 'When you come this way again you will find better roads,' was the remark that we heard all along the line."[228] And Henry Ford spoke his mind:

> The result of the race is just as I predicted at the start, and whereas I am greatly satisfied and pleased, I would have been very much more so had there been twenty entrants in the contest instead of five. The action of the A.A.A. in influencing the manufacturers in the contest as they have means that they have overlooked one of the greatest boons to the cause of motoring and good roads in the history of the automobile.[229]

Ford No. 1 Disqualified

Earle Chapin, the Shawmut crewman whose illness forced him to leave the car in Oregon, had taken the train ahead to witness the finish. "This has been a most remarkable and difficult run, and the Shawmut car has fought its way from the start in a manner which reflects credit on the drivers and the machine," Chapin said. "The Shawmut is a heavy car, and for days along the route it has ploughed its way through mud to the hubs. The car was in the lead at Fort Steel[e], Wyoming, but fell behind the two Fords sixteen hours while attempting to get a pass from the station agent to go over the river on the railroad bridge." As he added in a 1959 *Boston Sunday Globe* interview: "We couldn't have worked harder if we carried the car across the country on our backs."[230]

As winners, Scott and Smith would receive $2,000 in gold. Henry Ford had also decided to divide between Scott in Ford No. 2 and Kulick in Ford No. 1 "the money value of the Guggenheim trophy—$2,000," according to the *Chicago Daily Tribune*.[231] Guggenheim, however, refrained from awarding the prizes: the Shawmut Motor Company had notified race officials that it was preparing a formal protest.

Delayed four hours at the Lake Keechelus ferry landing, the Shawmut arrived on the expo grounds at 5:33 a.m. Thursday, June 24, "after an exhausting fight with deep snow in the Snoqualmie pass in the Cascades." (See Figs. 3.32 and 3.33.) Its time of 22 days, 17 hours,

Fig. 3.32 A mud-covered Shawmut on display with Ford No. 2. The rope running over the seats is an emergency support for the foot locker on the car's left side. (UWSC Nowell X2215)

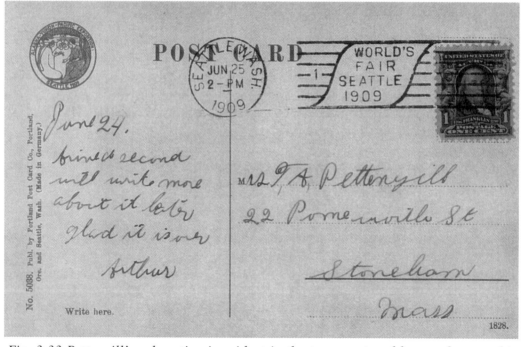

Fig. 3.33 Pettengill's exhaustion is evident in the terse postcard he sent home to his wife on the day the Shawmut arrived in Seattle: "finished second/will write more about it later/glad it is over." (SHS)

33 minutes put it in second place, 16 hours, 38 minutes behind the winning Ford. "Owing to the early hour of arrival the car was not greeted by the crowds. Driver P.A. [sic] Pettingill and Mechanic R.H. Messer, went to bed at the nearest hotel."[232]

Kulick and Harper in Ford No. 1 were due in Seattle late Thursday afternoon and also had sleep in mind. But on Snoqualmie Pass at 10 a.m. Thursday, "a rock hidden in the mud and snow sprang up to give us one last foul blow," recalls Harper, who nimbly avoids mentioning that his car broke an axle. "For seven hours we worked on the top of the mountain up among the clouds remedying the trouble that rock had caused."[233]

When the car hit the boulder, "the front axle snapped, and the transmission case was so badly bent the flywheel couldn't turn," Ford acknowledged in a 1959 chronicle of the race. Thus Kulick set to work "installing a new axle (which he knew would disqualify him) and hammering the dents out of the transmission case. They left the summit of the Pass at 5 p.m., traveled half a mile over the ties of the Milwaukee Railroad to the westward slope, still about 90 miles from Seattle."[234]

Ford No. 1 arrived in Seattle at 2:30 a.m. Friday, June 25, "but the drivers stopped at the Rainier-Grand, where they remained until 11:30 before going to the tape" at the exposition gate, finishing officially at 11:58 a.m., the *Seattle Post-Intelligencer* reported. Their official elapsed time was 23 days, 23 hours, 58 minutes. (See Table 3.3.) "However, Ford No. 1 will not be allowed third place. The crew changed the rear [sic] axle at Snoqualmie pass and thus forfeited the car's standing in the race," the newspaper reported.[235]

Table 3.3 Performance of Autos in 1909 New York–Seattle Race

Auto (place)	Elapsed time	Avg. daily mileage	Avg. speed
Ford No. 2 (disqualified)	22d-00h-55m	186	7.76 mph
Shawmut (Car No. 5) (1st place)	22d-17h-33m	181	7.53 mph
Ford No. 1 (disqualified)	23d-23h-58m	171	7.13 mph
Acme (Car No. 4) (2nd place)	28d-03h-24m	146	6.08 mph
Itala (Car No. 6) (withdrawn)	–	–	–
Stearns (Car No. 3) (withdrawn)	–	–	–

Acme, Shawmut Challenge Results

Finishing fourth, the Acme reached Seattle at 3:24 p.m. Tuesday, June 29, six days behind the winner. (See Table 3.4.) "It had continued in the contest in spite of much delay," the *New York*

Table 3.4 Daily Progress and Position of the 1909 Racers*

Date	Acme	Ford 1	Ford 2	Itala	Shawmut
6/1	Poughkeepsie, N.Y.-5	Poughkeepsie, N.Y.-1	Poughkeepsie, N.Y.-1	Poughkeepsie, N.Y.-4	Poughkeepsie, N.Y.-1
6/2	Syracuse, N.Y.-1	Syracuse, N.Y.-1	Syracuse, N.Y.-1	Syracuse, N.Y.-1	Syracuse, N.Y.-1
6/3	Buffalo, N.Y.	Buffalo, N.Y.	Buffalo, N.Y.	Buffalo, N.Y.	Buffalo, N.Y.
6/4	Toledo, Ohio-3	Toledo, Ohio-1	Toledo, Ohio-1	Toledo, Ohio-4	Toledo, Ohio-5
6/5	Chicago-3	Chicago-1	Chicago-1	Chicago-5	Chicago-3
6/6	Bloomington, Ill.-4	Springfield, Ill.-1	Springfield, Ill.-1	Chicago-5	Bloomington, Ill.-3
6/7	E Missouri	E Missouri	E Missouri	Litchfield, Ill.-5	E Missouri
6/8	E Kansas	E Kansas	E Kansas	Litchfield, Ill.-5	E Kansas
6/9	C Kansas	C Kansas	C Kansas	St. Louis-5	C Kansas
6/10	W Kansas	W Kansas	W Kansas	—	S Colby, Kan.-1
6/11	Quinter, Kan.-4	Burlington, Colo.-2	Burlington, Colo.-2	—	Burlington, Colo.-1
6/12	Oakley, Kan. ?-4	E Denver-2	Denver-1	—	E Denver-3
6/13	—	Greeley, Colo.-3	Cheyenne, Wyo.-1		Cheyenne, Wyo.-2
6/14	Pierce, Colo.-4	Medicine Bow, Wyo.-3	Rawlins, Wyo.-1	Marshall, Mo.-5	Fort Steele, Wyo.-2
6/15	N Colorado-4	Rawlins, Wyo.-2	Rawlins, Wyo.-2	—	Wamsutter, Wyo.-1
6/16	N Colorado-4	Opal, Wyo.-1	Opal, Wyo.-1		Opal, Wyo.-1
6/17	N Colorado-4	Pocatello, Idaho-1	W Cokeville, Wyo.-3	Salina, Kan.-5	Bancroft, Idaho-2
6/18	Bosler, Wyo.-4	E Shoshone, Idaho-1	W Pocatello, Idaho-2	—	American Falls, Idaho-3
6/19	Bosler, Wyo.-4	C Idaho-2/3	Boise, Idaho-1	—	W Buhl, Idaho-2/3
6/20	W Bosler, Wyo.-4	N Boise, Idaho-2	N Baker City, Ore.-1	Denver-5	N Boise, Idaho-3
6/21	W Rawlins, Wyo.-4	W Walla Walla, Wash.-3	E North Yakima, Wash.-1	—	W Walla Walla, Wash.-2
6/22	—	C Washington-3	Snoqualmie Pass, Wash.-1		E Ellensburg, Wash.-2
6/23	W Pocatello, Idaho-4	W Ellensburg, Wash.-3	Seattle-1	Cheyenne, Wyo.-5+	E Renton, Wash.-2
6/24	American Falls, Idaho-4	W Snoqualmie Pass, Wash.-3			Seattle-2
6/25	—				
6/26	ca. Walla Walla, Wash.-4				
6/27	—				
6/28	—				
6/29	Seattle-4				

* The racers often traveled all night. Therefore, their westward progress as of midnight on any given day may have been greater than reported in newspapers and auto journals, from which this information is drawn.

+ Withdrew from the race at Cheyenne.

Times said, politely.[236] Acme representative George Salzman, in Pocatello when Ford No. 2 reached Seattle, claimed the Shawmut and both Fords had disqualified themselves by violating the Automobile Club of America's rules for the race. According to the *Pocatello Tribune*, Salzman "asserts that a short distance out of St. Louis the Acme passed the Ford car No. 2 and discovered that it was being driven by a pilot, in violation of one of the rules." In Baker City, Salzman later claimed that both Fords disqualified themselves for replacing parts in Idaho— Ford No. 1 an axle at Twin Falls and Ford No. 2 an engine at Montpelier.

"In support of his contention Mr. Salzman produced affidavits from different towns where the cars stopped stating that parts had been replaced and other things done that were non-ethical, to say the least," according to a *Baker City Herald* article headlined "Ford Cars Are Under a Cloud." Salzman, in addition, "claims the Ford people bought up ferrymen [and] hired guards to watch bridges."[237] Like Ford No. 1, the Shawmut replaced a broken axle rather than weld the break, as the Acme crew did to its own broken axle, Salzman charged. "We have not broken a single rule of the game, and I confidently believe we will be declared winners when we reach Seattle."[238]

Shawmut Motor Company attorney James B. Howe received sworn statements from the Shawmut crew as soon as the men reached Seattle. Even before the second Ford arrived, Howe filed a protest against both Fords. The company's "principal charges" numbered five, according to reports in the *Seattle Post-Intelligencer*, *New York Times* and elsewhere:

1. Ford No. 2 "arrived in Seattle with a new axle, one that had not been stamped in New York City, thus breaking the rules against putting in new axles if the old one had worn out."
2. While Scott and Smith slept at the summit of Snoqualmie Pass in the Cascades, "an employe[e] of the Ford branch in Seattle met the racer, and proceeded to get into the driver's seat and operate the car. Later, the regular driver, who had sent the car across the continent, went back at the wheel."[239] Further, "It is charged that the Ford people arranged the change in crews."
3. "That at Lake Keechelus the Shawmut was held up four hours by the failure of the ferryman to appear, and it is stated elsewhere than in the official protest, that the failure of the ferryman was due to his having been 'seen' by someone anxious to get the Ford in first."
4. At Fort Steele, Wyoming, where the wagon bridge was washed out, "the two Ford cars were allowed to cross on the railroad bridge by virtue of special permits, while the Shawmut was refused access to the bridge, being delayed sixteen hours."
5. At Glasgow, Missouri, the Fords crossed the Missouri River "on a ferry which broke down before the Shawmut arrived, and the latter was obliged to cross on a railroad bridge."[240]

Guggenheim Hears Protest

Friday afternoon, after Ford No. 1 crossed the finish line, Guggenheim, the race referee, held a formal hearing of the charges. The crewmen from the Ford and Shawmut racers crowded into a room at the expo's Administration Building, along with Shawmut and Ford officials—including Henry Ford. The scene "very much resembled a trial in a suit for damages in the superior court," the *Seattle Post-Intelligencer* observed. There was one important difference, however, and A.J. Falknor, representing the Shawmut company, "was very much perturbed from the fact that the witnesses were not under oath."[241]

Guggenheim's only action was to disqualify third-place Ford No. 1. As Henry Ford told reporters after the meeting: "I will admit that Ford No. 1 was disqualified, and such statement, together with the reason, was given to the referee immediately upon its arrival, which will allow the Acme car third place." The rest of the hearing was simply that—a hearing of the protest. On the lesser charges involving the Shawmut's delays at Glasgow and Fort Steele, "the Ford cars won a good lead through what the Ford people call 'generalship'"—a reference to skillful management or tactics, as in war. But these "are matters which may not be influenced by the rules of the contest," said the *Seattle Post-Intelligencer*.[242]

If proven true, however, the allegations regarding the axle change and unauthorized driver "will certainly disqualify the Ford car which finished first," the newspaper added. In answer to the charges,

> Scott and Smith deny the allegations made in the complaint in regard to the changing of the axle or of sleeping while their car was put over the pass by others. They deny having at any time on the route conducted themselves toward the competing cars in a manner other than would be in keeping with clean sport. They cited several instances where they claim to have stopped and assisted the Shawmut car out of difficulty.[243]

Falknor requested permission to resume the hearing Monday afternoon, giving him time to secure witnesses to testify. But "Mr. Ford wanted to go bear hunting; Scott and Smith wanted to go to California, and others had several pursuits which are placed in the future by the delay," as the Seattle newspaper put it.[244] Guggenheim evidently denied the request and adjourned the hearing without announcing his decision on the protest.

Afterward, "the drivers of the Shawmut car sought the Ford men in the hall and avowed their feeling of friendship toward them, saying that the matter of the protest was not one of their inauguration." Henry Ford said:

> It is not necessary for me to say at length what I think of the Shawmut car. My testimonial rests in the fact that after the demonstration it has made across the continent I intend to buy one when I get back to Detroit.

I am very much surprised that the Shawmut should have entered the protest, for the winning car has in no way violated the rules governing the contest … I would under no circumstances accept the cup had my car been guilty as the protest alleges, but knowing, as I do, that the car came through with a "clean bill," I will not under any condition surrender the victory it has legitimately won.[245]

In a special "Trans-Continental Race Issue" of *Ford Times* magazine, the automaker airily dismissed the charges: "The decision was protested on a technicality. But that was baby play—a repetition of the old story of the boy who lost his marbles and wanted them back."[246]

Guggenheim Makes Decision; Committee Reverses It

In opening an awards banquet on Tuesday evening, June 29, Guggenheim announced he was denying the Shawmut company's protest "because its testimony did not establish any fraud, as charged, on the part of the Ford representatives," as *the Seattle Post-Intelligencer* paraphrased it. He also rejected a Ford counter-protest against the Shawmut "because it was not based on any statement of a violation of the rules, but merely upon the statement that … the Shawmut car had violated certain rules of the contest. No evidence was submitted."

Arriving just hours before Guggenheim's announcement, the Acme crewmen said that they, too, would file a formal protest. They evidently had no time to do so. Immediately after his announcement, Guggenheim awarded the trophy and $2,000 to Ford No. 2 and $1,500 to the Shawmut, and also announced that he had decided to hold the race annually. His decision in making the awards "was subject to revision by the Automobile Club of America should any formal protest be entered."[247]

The Shawmut entered a formal protest. The contest committee of the Automobile Club of America met three times—August 2, October 25 and October 28—to hear evidence, voting unanimously at the last meeting to overrule Guggenheim's decision, based on a new charge against the winning Ford. The committee's official report concluded:

It was proven to the satisfaction of the committee that Ford Car No. 2, in the New York to Seattle Contest, was guilty of a violation … in that it traveled a part of the distance between New York and Seattle with an engine which was substituted in the place of the engine stamped in New York by the technical committee of the Automobile Club of America and is therefore disqualified.

The decision of the referee in awarding first place in the contest to Ford Car No. 2 is hereby reversed and Shawmut Car No. 5 is declared the winner, and it is hereby directed that the M. Robert Guggenheim Trans-Continental Trophy and prize be awarded accordingly.[248]

"There were several allegations in the protest against the Ford car," *Motor Age* reported, "but the one on which the board acted was that the stamped engine had been changed at Copley, Ohio."[249] Robert Lee Morrell of New York City, chairman of the ACA contest committee, on October 29 sent a telegram to Seattle, announcing the reversal. In reporting the news, the next day's *Seattle Post-Intelligencer* contended that Ford No. 2 changed engines not in Ohio but in Idaho:

> The Shawmut people appealed from the referee's decision to the highest automobile tribunal in the country and proved to the satisfaction of the A.C.A. contest committee that the Ford crew substituted an engine at Weiser, Idaho, which was enough to disqualify the Ford car.

> After a hearing in Seattle, Referee Guggenheim ruled that the Ford car No. 2 was the winner of the race. He wired Morrell yesterday he was satisfied and also wired his congratulations to the Shawmut company. The trophy and first prize were expressed to the Shawmut factory last night.[250]

What became of the trophy—and whether it still survives—is unknown. It is not, however, in the Henry Ford Museum & Greenfield Village at Dearborn, Michigan, according to a computer search of the museum's trophy holdings.

"It was proved that the Ford changed an engine," *Automobile* concurred in an early 1910 review of the race. The Shawmut company, by 1910 either out of business or still attempting to reorganize, had produced few autos "and there were no spare parts to be had," the journal noted. "Under those circumstances, it was a very courageous piece of work to enter the hard race and, moreover, see it through."[251] Pettengill's sister, Grace Leacock, memorialized the Shawmut racer and its crew in verse. (See Fig. 3.34.)

"There was an argument about the motor—changing the motor in the car," Smith recalled in a 1951 interview, part of a Ford Motor Company oral-history project. "Well, they had the motor stamped, the head, the block, and everything. So we said, 'Well, take the prints; they're right there.' They sent them back and said there was nothing to the charges; we had the same engine all the way through."[252] According to Philip Van Doren Stern, who wrote a history of the Model T, Smith told an interviewer in 1954 "that the engine had never been changed to his knowledge, although he admitted that it might have been changed without his knowing it while the car was overhauled en route by one of the many Ford dealers who were eager to see it win."[253]

Ironically, the Ford company had insisted upon the rule under which both of its racers were disqualified. According to Harper in Ford No. 1:

> It was greatly through the influence of our company that the condition by which every part of the contesting cars were to be numbered, and the cars disqualified in case of replacing any one of them, was made a part of the agreements under which the contest

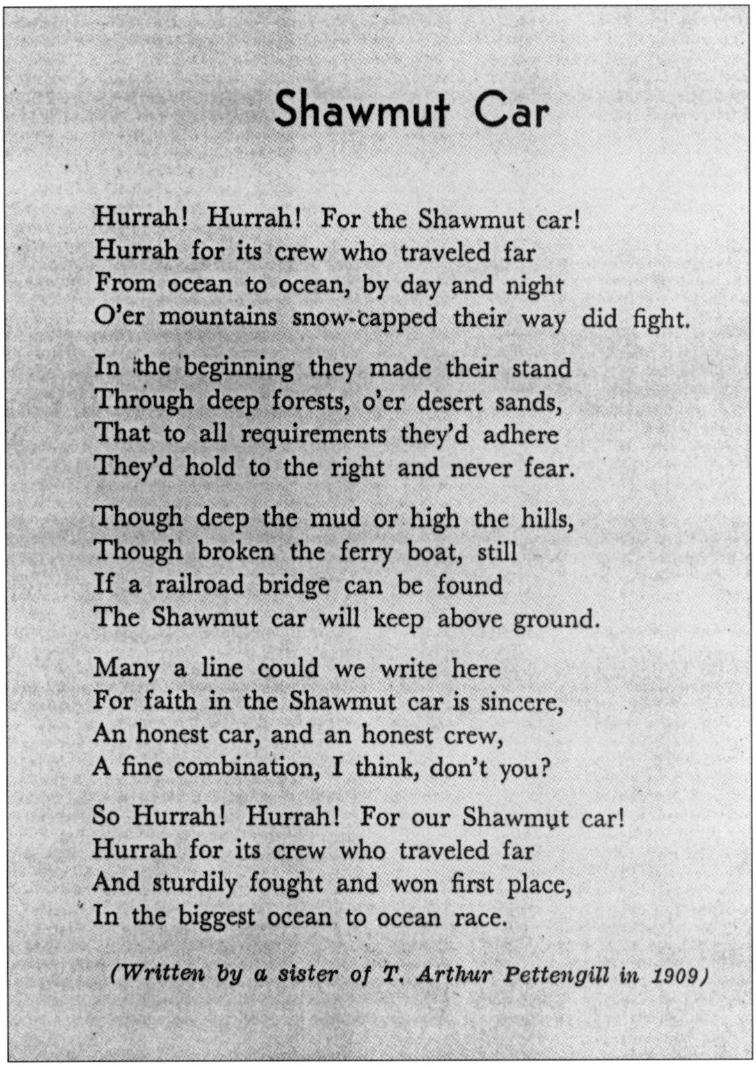

Shawmut Car

Hurrah! Hurrah! For the Shawmut car!
Hurrah for its crew who traveled far
From ocean to ocean, by day and night
O'er mountains snow-capped their way did fight.

In the beginning they made their stand
Through deep forests, o'er desert sands,
That to all requirements they'd adhere
They'd hold to the right and never fear.

Though deep the mud or high the hills,
Though broken the ferry boat, still
If a railroad bridge can be found
The Shawmut car will keep above ground.

Many a line could we write here
For faith in the Shawmut car is sincere,
An honest car, and an honest crew,
A fine combination, I think, don't you?

So Hurrah! Hurrah! For our Shawmut car!
Hurrah for its crew who traveled far
And sturdily fought and won first place,
In the biggest ocean to ocean race.

(Written by a sister of T. Arthur Pettengill in 1909)

Fig. 3.34 Pettengill's sister, Grace Leacock, wrote a poetic tribute to the transcontinental Shawmut.

was to be run. Punctures are not included, as they are likely to occur at any time, and do not affect the durability of any machine.[254]

Scott and Smith left Seattle July 9, drove Ford No. 2 to San Francisco and Los Angeles, and left there August 1 to begin a victory lap back to New York City, stopping at Ford agencies along the way. "No more ocean to ocean races for me," Scott told the *Kansas City Star*. "I have driven my last car in a race of that kind. It was bad going out, driving all day and working half the night and the bad roads have made the return trip almost as hard on us."[255] (See Fig. 3.35.)

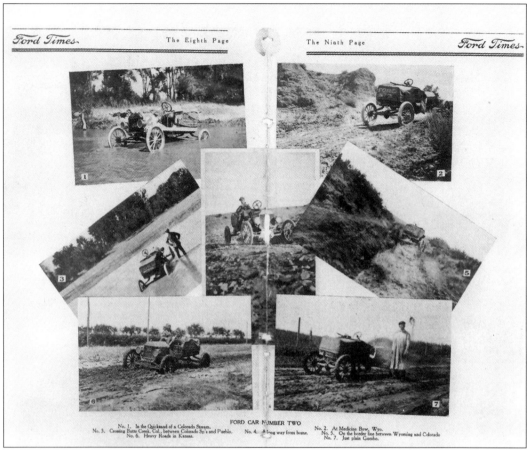

Fig. 3.35 A Ford Times *photo montage shows Ford No. 2 traveling some bad roads during the race and on its return trip. (Sept. 15, 1909,* Ford Times */HFM 0.3892)*

Then and Now

The 1909 transcontinental race accomplished the organizers' main goals by winning free publicity for the yearlong Seattle exposition and demonstrating the need for better roads. Guggenheim, despite forming a company to build his coveted New York–Seattle highway, was longer in seeing his wish fulfilled. Likewise, the coast-to-coast race did not become an annual event, despite Guggenheim's stated intentions.

Even though both Fords were disqualified, their performance gave an incalculable boost to the automaker's prestige. More enduring than any other result of the race, however, is the notion that Ford No. 2 won. Ford Motor Company has encouraged this revisionist view of history. Despite the disqualification of both its racers, the automaker marked the 50th anniversary of the 1909 New York–Seattle by retracing the route with a "caravan of Fords." The showpieces of the caravan: the company's 50 millionth car, a 1959 Galaxie, and a replica of the Ford No. 2.

According to a Ford press release, the caravan, a variety of antique autos and a "walk-through historical exhibit van" would demonstrate improvements to highways and autos during the previous half-century. Another purpose was to "remind a nation of the contributions which the automotive industry has made in the past and will make in the future to the American economy." With a bow to the future, the traveling caravan would also display Ford's experimental Levacar, "a wheel-less vehicle that travels on a film of air."[256]

The caravan left New York City June 1 and met with Ford agents and various city and state officials en route to arriving in Seattle on June 23, exactly 50 years after Ford No. 2 arrived there. William Clay Ford, the grandson of Henry Ford, met the caravan at the finish, on the site of Seattle's Century 21 Exposition, scheduled to open in 1961 and provide "a broad look into the world of the future."[257]

"None of Ford's press releases, nor a film of the expedition, released in 1962, made any mention of the disqualification of the original Ford 'Winner,'" David L. Lewis wrote in a 1971 article for *Ford Life*, a magazine about "Fords, People, Places and Events," published in Stockton, California. "Similarly, the Henry Ford Museum, which displays Ford No. 2 [a replica], ignores the disqualification. In describing 'winning No. 2,' the Museum's placard simply states that No. 2 'rolled into Seattle on June 23, 1909, 20 days (net) after leaving New York, and 17 hours ahead of the Shawmut. Third was the other Model T.'"[258]

The Henry Ford Museum & Greenfield Village in Dearborn, which no longer displays a replica racer, is now more truthful about the outcome. Today, even the computerized index in the museum's research center refers to the New York–Seattle race as one "in which a Ford finished first but was disqualified."[259]

The 50th anniversary celebration of the race presented the Ford Motor Company with an opportunity to reveal the fate of its two original racers. But a variety of articles in auto journals and newspapers, based on Ford press releases, neglect to do so. The two Fords, as well as the other New York–Seattle racers, are apparently lost to history.

In August 1945, Chapin, who had relocated to Reading, Pennsylvania, joined Messer and Pettengill of Stoneham at a Stoneham Rotary Club meeting. The *Stoneham Independent* played it up as a reunion for the three Shawmut drivers, 36 years after their victory in the 1909 race.[260] (See Fig. 3.36.) The article, however, contained no mention of the whereabouts of the Guggenheim trophy or the fate of the winning Shawmut auto. Messer, who became a mortician, died in Stoneham on October 4, 1952, at age 67.[261] Pettengill, who opened an auto-repair business in 1911 and later owned agencies for Dodge, Reo, Franklin and Nash autos, died in Stoneham at age 77 on March 3, 1955.[262] Chapin ran a garage in Stoneham and, after moving to Reading, operated a garage and started the Chapin Oil Company. He died on March 11, 1964, at age 81.[263]

Fig. 3.36 Left to right: Messer, Chapin and Pettengill during an August 1945 reunion in Stoneham, Massachusetts. (SHS)

In 1984, more than 30 Model T Fords left New York City for Seattle on a 25-day tour that the Model T Ford Club International organized to commemorate the 75th anniversary of the 1909 race.[264] Few participants seemed aware that the Shawmut actually won the original contest and the Acme placed second. Newspapers and auto journals covering the 1959 and 1984 ceremonial treks generally perpetuated the myth that a Ford had won the Great Race of 1909. For instance, the 1959 caravan would "mark Ford's victory in the grueling 1909 contest," the *New York Times* asserted.[265]

Three Races: An Historical Perspective

Regardless of how people remember it today, the 1909 race, coming just a year after the New York–Paris race crossed America, offered further proof of what an automobile could do. Four of six starters finished—and a fifth, the wayward Itala, might have done so had Guggenheim let it continue past Cheyenne. The winning car's 22-day crossing time halved the 44-day mark that Old Scout, the diminutive Oldsmobile, set in its race four years earlier, an indication of the greater power and speed that the newer autos possessed.

Thus, the New York–Seattle race ended a short but thrilling and instructive chapter in the history of the American automobile. The lack of support by various automotive organizations—among them the American Automobile Association, the Association of Licensed Automobile Manufacturers and the Manufacturers' Contest Association—rang down the curtain on coast-to-coast racing. Understandably, the various industry groups felt that the danger of an accident involving a contestant and another user of the public roads was too great to justify the risk of holding further races. Without the endorsement of these groups, automakers were generally reluctant to enter a transcontinental race.

What lessons did the 1905 Oldsmobile race, the 1909 New York–Seattle race and the dash across America by the 1908 New York–Paris racers teach, and how did these lessons benefit the production and sale of automobiles? One lesson of the three races came through loud and clear: the condition of rural roads in the United States was entirely at the mercy of the weather. In addition, the races demonstrated that many "roads" were little more than trails, devoid of improvements. Supposed improvements were often illusory. For instance, Old Steady broke through a Nebraska bridge in 1905; likewise, the Thomas broke through an Iowa bridge during the 1908 New York–Paris race. A year later, the New York-Seattle racers demonstrated again that bridges—flimsy or otherwise—were so often lacking that a coast-to-coast motor trip was anything but safe and comfortable.

Time and time again, the weather determined the pace of a race. When it rained, the racers crawled through mud and bumped through deep and dangerous washouts. Often, they had to push their cars or hire horses to tow them. When it snowed, the racers could hole up (though they generally didn't), shovel their way through deep drifts, or abandon the roads altogether in favor of farm fields or railroad tracks. Because news of the New York–Paris race was broadcast worldwide, the 1908 contest was particularly effective in shocking Americans into an awareness that they lacked "one of the best proofs of civilization—good roads."[266] Unfortunately for automobile owners, the three races, while focusing attention on America's primitive highway system, otherwise did little to bring about any immediate improvements.

Agitation for a coast-to-coast highway, however, began almost the moment the first car spanned the continent. In 1905, the U.S. Department of Agriculture's Office of Public Road Inquiry joined the Olds Motor Works in promoting the transcontinental race between two Oldsmobiles, a race undertaken to show the need for good roads. The 1909 New York–Seattle race shared that goal, and also strove to create a New York–Seattle coast-to-coast highway. Yet in the wake of the 1909 race, the automobile journals carried no news to suggest that Guggenheim's planned $100 million road-building corporation built so much as one mile of road.

In fact, the United States did not get its first transcontinental motor route until 1913, when a private association laid out and marked the Lincoln Highway between New York City and San Francisco. "Roads in the United States are said to be the worst of any civilized country in the world," observed Henry B. Joy, president of the Lincoln Highway Association. Yet its very name was too glamorous for the mud-slickened trail that the Lincoln Highway became after a rainstorm, according to one recreational motorist who drove that route in 1915.[267] "As is always the case," Joy added, "Congress and the State legislatures have been far behind the people in appreciating the need for highway development."[268] But this would change.

The burgeoning demand for good roads—stimulated in part by publicity surrounding the three transcontinental auto races and a number of other coast-to-coast treks—culminated in the 1916 passage of a federal-aid highway bill. Dedicating federal money to road building led to the gradual improvement of highways into the 1920s and beyond.

During the four-year period from 1905 to 1909, automobiles showed greater improvement than the roads they plied. The two 1904 Oldsmobile runabouts that staged the first transcontinental race in 1905 were little more than motor buggies, their 1-cylinder, 7-horsepower engines mounted amidships beneath the body. Though lightweight, their top speed was about 30 mph—fast enough for a tiller-steered car traveling on rough dirt roads.

As the only American entry in the 1908 New York–Paris race, the Thomas Flyer represented the progress made in the U.S. auto industry since 1904. During that short period, automobiles became longer, heavier and more powerful. The Thomas, for instance, used a 4-cylinder, 70-horsepower engine mounted under a hood at the front of the car, instead of positioned in the center of the frame beneath the body. At 4,250 pounds, the Thomas was more than five times heavier than an 800-pound Oldsmobile runabout. Despite being equipped with an outdated chain drive (instead of a drive shaft), the Thomas proved itself to be the mechanical peer of some of the best foreign automobiles.

The American autos entered in the 1909 New York–Seattle race demonstrated that two schools of thought existed regarding the car of the future. Neither school put on an especially convincing showing. Henry Ford claimed that the extensive use of vanadium steel in his Model T allowed him to produce a lightweight yet rugged automobile, superior to the heavier cars then in vogue. Regardless, Ford No. 2, weighing 950 pounds, arrived at the tape just 17 hours ahead of the 4,500-pound Shawmut. Had the Shawmut been able to avoid a delay of nearly that length at the Fort Steele, Wyoming, railroad trestle, it may have tied or even beaten the Ford.

Despite their differences in design and weight, both cars required regular stops to repair a variety of parts, including such major components as engines and axles. Yes, the American automobile industry had shown tremendous progress during its first decade; but there was plenty of room for improvement. Even so, the coast-to-coast racers proved that an automobile could survive some hard knocks. Or, in the words of Dwight B. Huss of the Old Scout crew: "You would be astonished at what an automobile will stand and still be in working order."[269]

One other important outcome of the 1905, 1908 and 1909 races was the simple notion that, despite the difficulties, it *was* possible to drive across the country. No longer did the railroads hold a monopoly on transcontinental travel. This encouraged average Americans to attempt the same feat in ordinary automobiles, not specially equipped racing cars.

"I had been reading in the trade papers with great interest … the stories of the present New York-to-Paris race," observed Jacob M. Murdock, a Pennsylvania lumberman.[270] Thus inspired, and while the New York–Paris race was still in progress, he set off from Los Angeles on April 24, 1908, with his wife, two daughters and a young son aboard the family's Packard touring car. Thirty-two days later, he reached New York City and entered the history books as the first American to drive his family on a coast-to-coast auto trek.

In 1911, a group of wealthy Americans confirmed Murdock's assertion that an average motorist could make the long journey a safe and enjoyable outing. Comfort was a goal for the 40 men, women and children who boarded 12 Premier autos for a sightseeing trip from Atlantic City, New Jersey, to Los Angeles. "There is a general feeling that the Pacific and Atlantic coasts have been brought closer together," *Motor Age* observed following the run. Transcontinental touring for pleasure would become common because of the Premier tour's "unqualified success," the magazine predicted. Because of the journey, the Premier Motor Manufacturing Company decided to produce a "Premier de luxe," designed especially for long-distance travel.[271]

Taking a cue from the popularity of the three races, a number of U.S. manufacturers later sought to promote their automobiles by sponsoring one-car transcontinental trips. Thus, in 1915 and 1916, at least seven autos set coast-to-coast speed records. When America's entry into World War I interrupted such stunts, a Hudson held the crossing record with a time of 5 days, 3 hours, 31 minutes, which rivaled the speed of the fastest trains and represented just 12 percent of Old Scout's 1905 crossing time.

As to the production and sale of automobiles, excitement surrounding the three transcontinental races provoked a great public interest in the self-propelled machines. To many Americans who read daily accounts of the races, the automobile proved its utility in a dramatic way. Midwestern farmers rode horses for miles to get a glimpse of the 1905 Oldsmobile racers, noted Old Scout pilot Dwight B. Huss. They had a "downright practical" interest in what the automobile could do for them, he noted. "If it was to be worthwhile for them it had to be improved transportation, not a playtoy."[272] A Thomas agent who observed the same Midwestern phenomenon three years later declared that the 1908 New York–Paris race "has done more to interest them in automobiles than all the [automobile] shows." Spectators did more than just watch the round-the-world racers pass by, one Midwestern daily newspaper noted. "Dealers say that their sales have increased mightily and that they are constantly besieged with prospective buyers asking for demonstrations."[273]

Individual automakers sought to capitalize on this excitement. Following the 1905 Oldsmobile race, Olds Motor Works advertised that, because its car was sturdy enough to cross the continent, "you will get pretty satisfactory returns for daily use."[274] Likewise, makers of the Thomas (in 1908) and Ford (in 1909) published promotional booklets heralding the performance of their vehicles.

Other hints suggest that the races inspired not only an appreciation for but an awe of the automobile. Veteran racer Montague "Monty" Roberts declared the New York-Paris race to be "the most popular competition ever held in automobiling."[275] Autograph seekers and spectators with cameras besieged Bert Scott and Jimmy Smith when their Ford No. 2 crossed the finish line in the 1909 New York–Seattle race. Smith became so widely recognized that his photo pasted on an envelope and addressed to "Ford Motor Company, Detroit, Michigan" was all that was necessary to get a letter to him even months after the race.

For the United States, the first decade of the twentieth century was an age of discovery and conquest. Under swashbuckling President Teddy Roosevelt (1901–1909), America set out to tame the Trusts—Big Oil, Steel, Railroads, and other monopolies—and undertook another gargantuan task: digging the Panama Canal. Orville and Wilbur Wright ushered in the era of modern aviation by flying 120 feet in a powered heavier-than-air machine on December 17, 1903. That same year, Dr. H. Nelson Jackson and Eugene I. Hammond became the first persons to drive an automobile from coast to coast. During the first decade of the 1900s, Americans explored the Arctic and Antarctic; Edward Payson Weston walked across the continent (twice); and the "Great White Fleet" cruised around the world to demonstrate the might of the U.S. Navy.

Against this backdrop occurred three of the most exciting automobile races ever conceived—races that spanned the continent and tested man and machine alike. Driving skill was only one prerequisite of winning. More important to the outcome was mechanical ability, preparation, determination, resourcefulness and teamwork. These races, occurring while the automobile was still quite young (and, in the minds of many, unproven), captivated a nation and established the dominant transportation form of the twentieth century and beyond. A modern automobile race around an asphalt track may have its exciting moments. But such excitement pales in comparison to the physical and mental demands of a transcontinental race, in which many of the dangers were unknown, battling the elements was as important as beating the other contestants, and the most elemental goal was simply to finish in one piece. From our vantage point in an age of paved turnpikes and comfortable, air-conditioned cars, it is only natural that the incidents of three early transcontinental automobile races have taken on epic proportions.

Notes

Notes to Chapter 1

1. "'Old Scout' Drives In," *Cheyenne (Wyo.) Daily Leader*, June 1, 1905, p. 8:3.
2. One of the automaker's post-race ads, "'Old Scout' Wins the Transcontinental Race," *Automobile Topics*, undated, p. 892, refers to the only change in the cars' standard equipment "consisting in enlarged gasoline and water capacity." It is unclear whether Olds Motor Works was alluding to larger tanks being installed before the race or to the extra tanks that both cars installed in the Midwest.
3. "Fast Run to Albany," *Automobile*, May 13, 1905, p. 595:2.
4. From an untitled early Olds owner's manual, reproduced as a 24-page booklet "and distributed by the Public Relations Dept. of Oldsmobile Division, General Motors Corporation." Though a note identifies it as a 1904 owner's manual, p. 7 refers to "the Oldsmobile Transmission gear for 1903...."
5. "Origins of Old Scout," *Car Classics*, April 1972, n.p., quotes John Hammond II as saying the Whitman-Hammond 1903 Olds transcontinental car became Old Scout: "Mr. [Ransom E.] Olds told my dad and me that he had Whitman drive the same car in the New York to Pittsburg[h] Endurance Run; then it was shipped back to the factory where it was 'modernized' as a 1904 model, replacing the wire wheels with wooden-spoked ones, etc. It was then christened 'Old Scout' and was one of the two cars driven from New York to Portland, Oregon, in the 'race' to the Lewis and Clark Exposition." Hammond II evidently couldn't support this assertion; he makes no mention of it in his book, *From Sea to Sea in 1903 in a Curved Dash Oldsmobile* (Egg Harbor City, N.J.: Laureate Press, 1985), 166 pp.
6. "Olds European Tour Ends," *Automobile*, Jan. 7, 1905, p. 12:1; and Olds ad, "$1,000 Prize/Oldsmobile Volunteers Wanted," *Automobile Topics*, undated copy, p. 151.
7. "Route to be Followed West of Omaha," *Automobile*, May 25, 1905, p. 645:1.
8. "Hot Fight for President," *Portland (Ore.) Morning Oregonian*, June 22, 1905, p. 16:2. Abbott apparently left his post shortly after the 1905 race. *Automobile* described Abbott, author of a three-article race summary, as "formerly special agent, U.S. Office of Public Roads."
9. "With the Overland Car through Utah into Colorado," *Horseless Age*, Sept. 30, 1903, p. 359:2. Abbott's special interest was in the roads themselves, and how to improve them. Thus he wrote "Use of Mineral Oil in Road Improvement," *Yearbook of the Department of Agriculture, 1902* (Washington, D.C.: Government Printing Office, 1903), pp. 439-54.
10. "'Old Steady' Pair Here," *Chicago Daily News*, May 15, 1905, p. 2:2 of "Late Sporting Edition"; and "Ocean to Ocean Race," *New York Tribune*, May 9, 1905, p. 5:6.

11. Cy Linder, "Transcontinental Auto Tourists Go through Chicago," *Chicago Inter Ocean*, May 16, 1905, p. 5:1.

12. "Prizes for Ocean to Ocean Race Sometime Next Year," *Automobile*, Oct. 31, 1903, p. 467:2.

13. "Autos Start To-Day for Pacific Coast," *New York Times*, May 8, 1905, p. 11:3.

14. James W. Abbott in a series titled "Concerning the Recent Transcontinental Race from New York to Portland, Oregon," *Automobile*, Aug. 24, 1905, p. 210:1, and Aug. 10, 1905, p. 150:3.

15. "Start of Olds Transcontinental Race to Lewis and Clark Centennial Exposition," *Automobile*, May 13, 1905, p. 594:3.

16. *Automobile*, May 13, 1905, p. 594:1; and "Autos Off for Oregon," *New York Times*, May 9, 1905, p. 6:3.

17. "The Oldsmobiles Are Off for Portland," *Motor World*, May 11, 1905, p. 299:2.

18. *Automobile*, May 13, 1905, p. 594:1; and *New York Tribune*, May 9, 1905, p. 5:6.

19. James W. Abbott, "Concerning the Recent Transcontinental Race from New York to Portland, Oregon," *Automobile*, Sept. 7, 1905, pp. 259–60.

20. "On the Trail of Lewis and Clark," *Automobile Topics*, May 13, 1905, p. 316, says the opposite—Old Scout used Diamonds, Old Steady Fisk.

21. Percy F. Megargel, "Diary of the Transcontinental Race," *Automobile*, May 18, 1905, p. 622:2.

22. Reprinted early Oldsmobile owner's manual, p. 24.

23. *Automobile*, May 13, 1905, p. 594:3.

24. George S. May, *R.E. Olds: Auto Industry Pioneer* (Grand Rapids, Mich.: William B. Eerdmans Publishing Co., 1977), pp. 226, 228.

25. Information about Megargel's bicycling background comes from a May 14, 1905, *Nashville (Tenn.) American* article in a scrapbook about the 1905 race that is held by the Oldsmobile History Center, Lansing, Mich. Edd Whitaker of Portland, Oregon, producer-reporter of a documentary videotape "From Hellgate to Portland," transcribed many of the scrapbook articles and generously provided the copies used in this chapter.

26. For *Automobile*, Megargel wrote a series, "On the Road to St. Louis," the first five parts of which were in the issues of May 28, 1904, pp. 572–73; June 4, 1904, pp. 600–01; June 11, 1904, p. 631; June 18, 1904, pp. 661–62; and June 23, 1904, pp. 685–86.

27. As related in "Syracuse to Rochester," *Horseless Age*, Aug. 3, 1904, p. 109:1.

28. These biographical details are from two sources: *Portrait and Biographical Record of Lackawanna County, Pennsylvania* (New York: Chapman Publishing Company, 1897), p. 827:1; and *Automobile*, May 13, 1905, p. 595:1.

29. Robert Wallace, "Meet Me (Pop!) in (Crash!) St. Louis, (Squawk!) Louis," *Life*, July 12, 1954, p. 105:1; and *Automobile*, May 13, 1905, p. 594:3.

30. Edward M. Miller, "Trek of Old Scout over Cross Country Trail Automotive Epic," *Portland (Ore.) Sunday Oregonian,* Sept. 27, 1931, IV, p. 1:3.

31. "At Syracuse," *Horseless Age*, Aug. 3, 1904, p. 107:2.

32. *New York Times*, May 8, 1905, p. 11:3.

33. The average annual wage in all industries, including farming, was $503 in 1905, according to the U.S. Department of Commerce, *Historical Statistics of the United States: Colonial Times to 1957* (Washington, D.C.: U.S. Government Printing Office, 1960), p. 91.

34. "Cross-Continent Route," *Automobile*, May 6, 1905, p. 579:2.

35. Percy F. Megargel, "Diary of the Transcontinental Race," *Automobile*, June 29, 1905, p. 781:1.

36. *Automobile*, May 13, 1906, p. 595:3.

37. *Automobile*, May 18, 1905, p. 622:1.

38. *Automobile*, May 18, 1905, p. 622:2.

39. *Automobile*, May 18, 1905, p. 622:2.

40. "Oldsmobiles Are Making Fast Time," *Erie (Pa.) Dispatch*, May 12, 1905, p. 8:2.

41. "Will Not Drive in Cup Race … Oldsmobile Transcontinental Drivers Make Good Time over Bad Roads," *Cleveland Plain Dealer*, May 12, 1905, p. 10:5.

42. "Cross-Continent Cars Off in Mud," *Cleveland Leader*, May 13, 1905, p. 8:6.

43. "Oldsmobile Transcontinental Race," *Cycle and Automobile Trade Journal*, June 1, 1905, p. 54:2.

44. "Auto Race across Continent," *Toledo (Ohio) Blade*, May 13, 1905, p. 1:3.

45. *Cycle and Automobile Trade Journal*, June 1, 1905, p. 54:2.

46. *Toledo (Ohio) Blade*, May 13, 1905, p. 1:2.

47. *Automobile*, May 18, 1905, p. 622:3.

48. "Wallowing through Mud," *Motor World*, May 18, 1905, p. 341:3.

49. *Chicago Daily News*, May 15, 1905, p. 4:3 of "Late Sporting Edition."

50. *Automobile*, May 18, 1905, p. 622:3.

51. Percy F. Megargel, "Diary of the Transcontinental Race," *Automobile*, May 25, 1905, p. 644:2.

52. *Automobile*, May 18, 1905, p. 622:3.

53. *Automobile*, May 25, 1905, p. 643:3.

54. *Cycle and Automobile Trade Journal*, June 1, 1905, p. 55:1.

55. Dwight B. Huss, "How Huss Tells It," in *From Hell Gate to Portland: The Story of the Race across the American Continent in Oldsmobile Runabouts, Told by the Men Who Rode and the Man Who Looked on* (Lansing, Mich.: Olds Motor Works, 1905), pp. 14–15. The title comes from the so-called Hell Gate area of New York City, where the transcontinental trip originated.

56. *Automobile*, May 25, 1905, p. 644:2.

57. *Automobile*, Aug. 10, 1905, p. 152.

58. *Automobile*, May 25, 1905, p. 644:3.

59. "Into the Western Wilds," *Motor World*, May 25, 1905, p. 388:3.

60. *Automobile*, May 25, 1905, p. 643:2, and Percy F. Megargel, "Diary of the Transcontinental Race," *Automobile*, June 1, 1905, p. 674:2.

61. From a 25-page unpublished typewritten history of Dwight Huss' experiences, *Chronology of [the] 1905 Olds Transcontinental Race*, p. 7, shared by Dwight's son, John Huss.

62. *Automobile*, June 1, 1905, p. 674:1.

63. Powell Motor Company ad, *Omaha (Neb.) Daily Bee*, May 24, 1905, p. 3:6.
64. *Automobile*, May 13, 1905, p. 594:2.
65. Olds ad, *MoTor*, July 1905, p. 15:1.
66. *Automobile*, June 1, 1905, p. 674:3.
67. Percy F. Megargel, "Diary of the Transcontinental Race," *Automobile*, June 15, 1905, p. 725:2.
68. *From Hell Gate to Portland*, p. 15.
69. *Motor World*, May 25, 1905, p. 388:3.
70. "Touring Cars on Long Trip," *Fremont (Neb.) Evening Tribune*, May 26, 1905, p. 6:1.
71. But, actually, Old Scout "went through the city last night," reported the next day's *Grand Island Daily Independent* in an untitled blurb, May 27, 1905, p. 8:5.
72. "Huss Shakes Off Megargel," *Motor World*, June 1, 1905, p. 432:2.
73. *Automobile*, June 15, 1905, p. 725:1.
74. "Old Steady Finishes Race," *Portland (Ore.) Evening Telegram*, June 29, 1905, p. 10:2.
75. "'Old Scout' Arrives," *Kearney (Neb.) Daily Hub*, May 27, 1905, p. 3:2.
76. "Oldsmobiles Coming," wire story from Kearney, Neb., in *Wyoming Semi-Weekly Tribune* (Cheyenne), June 2, 1905, p. 6:4.
77. "Huss Leads Olds Racers," *Automobile Topics*, June 3, 1905, p. 562.
78. "'Old Steady' Arrives," *Kearney (Neb.) Daily Hub*, May 29, 1905, p. 3:2.
79. "Autos on Long Trip," *Sidney (Neb.) Telegraph*, June 3, 1905, p. 1:2.
80. *From Hell Gate to Portland*, p. 15.
81. *Cheyenne (Wyo.) Daily Leader*, June 1, 1905, p. 8:3. The *Laramie Republican's* story, "'Old Scout' is Headed for Here," June 1, 1905, p. 1:1, said five escort cars left Cheyenne for Archer but only Tom Myotte's runabout made it there. The four remaining cars "were inadequate to get through the muddy roads." Three drove back to Cheyenne and the fourth, Dr. A.W. Crooks' auto, "stalled in the mud and was dragged in by Myotte's machine last night." (The doctor's name appeared as "Crook" in the *Cheyenne Daily Leader*, "Crooks" in the *Laramie Republican*.)
82. *Cheyenne (Wyo.) Daily Leader*, June 1, 1905, p. 8:3; and "Autos Coming," *Wyoming Semi-Weekly Tribune* (Cheyenne), June 2, 1905, p. 4:5.
83. *Wyoming Semi-Weekly Tribune* (Cheyenne), June 2, 1905, p. 4:5.
84. "Huss over the Rockies," *Motor World*, June 8, 1905, p. 480:1.
85. "'Old Steady' Reaches City," *Cheyenne (Wyo.) Daily Leader*, June 2, 1905, p. 8:2. Old Steady's accident was actually closer to Grand Island than to Columbus, Neb.
86. *Cheyenne (Wyo.) Daily Leader*, June 2, 1905, p. 8:2.
87. "First of Oldsmobiles Reaches the City," *Laramie (Wyo.) Republican*, June 2, 1905, p. 1:5.
88. *From Hell Gate to Portland*, p. 16.
89. *Motor World*, June 8, 1905, p. 480:1.
90. *Chronology of [the] 1905 Olds Transcontinental Race*, p. 17.
91. "'Old Steady' Makes Gain," *Cheyenne (Wyo.) Daily Leader*, June 4, 1905, p. 3:2.
92. *Automobile*, June 29, 1905, p. 780:1.
93. *Automobile*, June 29, 1905, pp. 780–81.

94. *From Hell Gate to Portland*, pp. 16–17.

95. *From Hell Gate to Portland*, p. 17.

96. *From Hell Gate to Portland*, p. 17.

97. "On Last Stage of Great Trip," *Idaho Daily Statesman* (Boise), June 12, 1905, p. 2:3.

98. Percy F. Megargel and Grace Sartwell Mason, *The Car and the Lady* (New York: Baker and Taylor Co., 1908), pp. 29–30. It's unclear why Mason and Megargel's book refers to Goodrich tires; Megargel used Diamond tires both on his 1905 Olds trip and on his later coast-to-coast journey in a Reo automobile.

99. "Auto Racers Reach Boise," *Boise (Idaho) Evening Capital News*, June 12, 1905, p. 8:1.

100. *Idaho Daily Statesman* (Boise), June 12, 1905, p. 2:3.

101. *Automobile*, June 29, 1905, p. 780:1.

102. *Automobile*, June 29, 1905, p. 781:1.

103. "Automobile Notes of Interest," *New York Times*, June 17, 1905, p. 6:4; and "Auto Paused on the Brink," *Detroit Evening News*, July 1, 1905, p. 4:3.

104. "Huss Six Days Ahead," *Motor World*, June 15, 1905, p. 521:1; and "Huss is Due Next Monday," *Portland (Ore.) Evening Telegram*, June 17, 1905, p. 8:1.

105. *Automobile*, Aug. 10, 1905, p. 152:1.

106. "Auto Leaves," *Idaho Daily Statesman* (Boise), June 14, 1905, p. 3:3.

107. "Old Scout Passes through the City," *Crook County Journal* (Prineville, Ore.), June 22, 1905, p. 1:6.

108. *From Hell Gate to Portland*, p. 19.

109. Background on the history and restoration of the Santiam Wagon Road comes from the author's July 6, 1995, interview with two U.S. Forest Service officials based in Sweet Home, Ore.—archaeologist Tony Farque and engineer Wayne Shilts. Another source was Percy Bush, "New Life for the Old Santiam Wagon Road," *Sweet Home (Ore.) New Era*, July 5, 1995, p. 13:1.

110. Dwight B. Huss, "Adventures of 'Old Scout' in Oregon, 1905," *Oregon Motorist*, September 1931, p. 4:1.

111. *Oregon Motorist*, September 1931, p. 4:1. This account is convoluted, for the men would have climbed Sand Mountain before Iron Mountain.

112. *Detroit Evening News*, July 1, 1905, p. 4:3.

113. *From Hell Gate to Portland*, p. 19.

114. "Old Scout at Salem," *Portland (Ore.) Morning Oregonian*, June 21, 1905, p. 6:1.

115. "Exactly two minutes to one," Huss specifies in *From Hell Gate to Portland*, p. 19.

116. "Huss Wins Race," *Automobile*, June 29, 1905, p. 782:1; and "'Old Scout' Wins Race," *Portland (Ore.) Morning Oregonian*, June 22, 1905, p. 16:4.

117. *From Hell Gate to Portland*, pp. 19–20.

118. "Good Roads Men Get Together," *Portland (Ore.) Morning Oregonian*, June 22, 1905, p. 16:1.

119. *Portland (Ore.) Morning Oregonian*, June 22, 1905, p. 16:4.

120. *Oregon Motorist*, September 1931, p. 5:2.

121. "Huss, Lionized at Portland, Talks of His Long Race," *Motor World*, June 29, 1905, p. 607:3.

122. *Detroit Evening News*, July 1, 1905, p. 4:3.
123. Percy F. Megargel, "Diary of the Transcontinental Race," *Automobile*, July 13, 1905, p. 47:2.
124. *Automobile*, July 13, 1905, p. 47:3.
125. "Second Auto in the Cross-Continent Race Reaches Portland," *Portland (Ore.) Morning Oregonian*, June 29, 1905, p. 10:5.
126. *Automobile*, July 13, 1905, p. 47:2.
127. *Automobile*, June 29, 1905, p. 782:3.
128. *Portland (Ore.) Morning Oregonian*, June 29, 1905, p. 10:5.
129. *Automobile*, June 29, 1905, p. 782:2; from an Old Scout fact sheet in the collection of the Oldsmobile History Center, Lansing, Mich.; and reprinted early Oldsmobile owner's manual, p. 21.
130. *Automobile*, July 13, 1905, p. 47:3.
131. *From Hell Gate to Portland*, p. 4.
132. *Portland (Ore.) Morning Oregonian*, June 22, 1905, p. 16:1.
133. *Motor World*, May 25, 1905, p. 388:3.
134. *Portland (Ore.) Morning Oregonian*, June 21, 1905, p. 6:1.
135. "Local Mention," *Crook County Journal* (Prineville, Ore.), June 29, 1905, p. 3:2.
136. "Oldsmobile Model S, 200-Hour Non-Stop at Detroit," photo and cutline in *Automobile*, May 17, 1906, p. 812:2.
137. "How They Ran and Why They Were Penalized," *Automobile*, July 18, 1907, p. 84:1.
138. "Eight Perfect in Ohio's 3-Day Reliability," *Motor Age*, Sept. 17, 1908, 13:1.
139. *Portland (Ore.) Morning Oregonian*, Sept. 18, 1931.
140. From an undated copy of an Oldsmobile ad titled "The Transcontinental Good Roads Tour."
141. From a speech John Huss gave July 5, 1995, at an Ontario, Ore., banquet commemorating the 90th anniversary of the 1905 Oldsmobile race.
142. *Life*, July 12, 1954, p. 108:2; and "Dwight Huss, Clyde Native and Early Auto Driver, Dead at 90," *Clyde (Ohio) Enterprise*, Sept. 3, 1964, p. 3.
143. "Early Death of Percy Megargel," *Automobile*, May 13, 1909, p. 786:1; and "Percy Megargel Dead," *Motor Age*, May 13, 1909, p. 11:3.
144. "News and Trade Miscellany," *Automobile*, Sept. 7, 1905, p. 276:3.
145. Malcolm Fisher, "'Old Scout' and 'Old Steady,'" *Antique Automobile*, March-April 1976, p. 31:1.
146. "Some Famous Oldsmobiles Which Helped Make Automotive History," *Oldsmobile Pacemaker*, January 1923, p. 6:2.
147. *Portland (Ore.) Sunday Oregonian*, Sept. 27, 1931, IV, p. 1:3.
148. An editor's note to *Oregon Motorist*, September 1931, pp. 3–5, also outlines the sale of Old Scout to Wemme and, later, to Cohen.
149. Malcolm W. Bingay, "Good Morning," *Detroit Free Press*, May 20, 1946, n.p.
150. An article, "Proud Old Scout Rides in Comfort for 2nd Transcontinental Journey," in an Oldsmobile employee magazine, *Team*, Summer 1975, n.p., pictures Old Scout being loaded onto the train. Under the heading "Railroads, History," the *New York Times Index* of 1975 and 1976 cites numerous articles about the American Freedom Train.

151. From telephone interviews with the author, Earley on May 28, 1993, and Hoonsbeen on June 23, 1993.

152. From the author's July 6, 1995, interview with archaeologist Tony Farque and engineer Wayne Shilts, U.S. Forest Service officials based in Sweet Home, Ore.

153. *Automobile*, Aug. 10, 1905, p. 150:1.

154. *Chicago Evening Post*, May 18, 1905, from a scrapbook held by the Oldsmobile History Center, Lansing, Mich.

155. Curt McConnell, "Nebraska Auto-Registration Figures for 1905–07," Society of Automotive Historians' *SAH Journal*, September–October 1995, p. 6:1.

Notes to Chapter 2

1. "Making Ready to Race from New York to Paris," *New York Times*, Dec. 29, 1907, V, p. 1:2.

2. "Across Three Continents by Automobile," *Automobile*, Dec. 5, 1907, p. 858:1.

3. "Steamers May Take Autos to Skagway," *New York Times*, Dec. 1, 1907, II, p. 1:3.

4. "Autoists Commend Paris Race Route," *New York Times*, Jan. 19, 1908, IV, p. 1:1.

5. Antonio Scarfoglio, trans. by J. Parker Heyes, *Round the World in a Motor-Car* (London: Grant Richards, 1909), p. 157.

6. *New York Times*, "When the Racing Autoists Arrive in Siberia," Feb. 23, 1908, V, p. 7:1., and Dec. 1, 1907, II, p. 1:3.

7. "Plans Announced for Race to Paris," *New York Times*, Jan. 5, 1908, II, p. 1:7.

8. *New York Times*, Jan. 5, 1908, II, p. 1:7.

9. "'Feasible,' Says Megargel," *New York Times*, Jan. 8, 1908, p. 7:3.

10. "Automobiles Ready for Their Great Race," *New York Times*, Feb. 9, 1908, V, p. 6:5.

11. "Finds Auto Route in Alaska Feasible," *New York Times*, Feb. 3, 1908, p. 5:2.

12. "Bets He'll Reach Paris in 100 Days," *New York Times*, Jan. 14, 1908, p. 3:3.

13. *New York Times*, Jan. 5, 1908, II, p. 2:1.

14. "Novel Devices for Polar Auto Race," *New York Times*, Dec. 23, 1907, p. 1:3.

15. "Run to Chicago to Take 5 Days," *New York Times*, Jan. 20, 1908, p. 2:2.

16. "How the Race Was Organized," *New York Times*, Feb. 13, 1908, p. 4:5.

17. "Preparing Autos for Polar Race," *New York Times*, Jan. 7, 1908, p. 4:1.

18. For the 13 of the 15 drivers or relief drivers whose ages were reported.

19. Quoted in Allen Andrews, *The Mad Motorists: The Great Peking–Paris Race of '07* (Philadelphia: J.B. Lippincott Co., 1965), p. 14.

20. "Foreign Cars Are Landed," *New York Times*, Feb. 11, 1908, p. 1:7.

21. *New York Times*, "Auto Racers May Start in February," Dec. 2, 1907, p. 1:1, "How the Foreign Cars in the New York to Paris Race Will Be Equipped," Feb. 2, 1908, IV, p. 1:2, and "Hansen Sure of Success," Jan. 5, 1908, II, p. 2:1.

22. "Racers Are Well Equipped," *New York Times*, Feb. 11, 1908, p. 2:5.

23. "Trying Out New York-to-Paris Motor Cars," *New York Times*, Jan. 26, 1908, V, p. 8:1.

24. *New York Times*, Feb. 2, 1908, IV, p. 1:3, and Jan. 5, 1908, II, p. 2:1.

25. Contradicting itself, the *New York Times* also referred to a 160-gallon tank of seven compartments.

26. *New York Times*, "Autoists Entered in Race to Paris," Feb. 9, 1908, IV, p. 1:1, "From Paris to New York," Feb. 9, 1908, II, p. 2:1, and "Foreign Autoists Entered in Race," Feb. 2, 1908, IV, p. 1:1.

27. *New York Times*, Feb. 9, 1908, IV, p. 1:1, and Jan. 14, 1908, p. 3:3.

28. Jeannette Mirsky, *To the Arctic! The Story of Northern Exploration from Earliest Times to the Present* (New York: Alfred A. Knopf, 1948), p. 310.

29. "Autoists Here to Begin Trip Back to Paris," *New York World*, Feb. 9, 1908, p. 3:6.

30. "American Car Gets a Great Reception," *Omaha (Neb.) World-Herald*, March 5, 1908, p. 5:3.

31. George N. Schuster and Tom Mahoney, *The Longest Auto Race* (New York: John Day Co., 1966), p. 28. The book expands upon an article Schuster wrote for the January 1963 *Reader's Digest*, "Around the World, Almost, in 169 Days," pp. 188–98.

32. Untitled dispatch, *New York Times*, Feb. 17, 1908, p. 2:3.

33. *New York World*, Feb. 9, 1908, p. 3:7.

34. *New York Times*, Feb. 2, 1908, IV, p. 1:3.

35. T.R. Nicholson, *The Wild Roads: The Story of Transcontinental Motoring* (New York: Norton, 1969), p. 142. A 17-year-old *New York Times* photographer spent some time traveling with the Thomas, but he wasn't an official crewman.

36. *Mad Motorists*, p. 6.

37. This summary of Godard's experiences is based on Andrews' account in *The Mad Motorists*.

38. *New York Times*, Feb. 9, 1908, II, p. 2:1.

39. *New York Times*, Feb. 2, 1908, IV, p. 1:5.

40. "Autoists Here for Big Race," *New York Times*, Feb. 8, 1908, p. 2:4.

41. "German Car Ready for World's Race," *New York Times*, Jan. 26, 1908, IV, p. 1:7.

42. *New York Times*, Feb. 8, 1908, p. 1:1, and "Two Autos Start for the Paris Race," Jan. 27, 1908, p. 1:5.

43. This account of Pons' experiences in the Peking-Paris race is based upon *The Mad Motorists*, pp. 136-37, 143.

44. *New York Times*, Feb. 9, 1908, IV, p. 1:2.

45. *The Wild Roads*, p. 143; and an untitled blurb in the *New York Times*, Feb. 15, 1908, p. 2:4.

46. Its color was in the eye of the beholder. The Thomas company's post-race publicity booklet, *The Story of the New York to Paris Race*, called it blue; three of the car's drivers called it gray. It was evidently a very light shade of gray, for the March 6, 1908, *Omaha (Neb.) Daily Bee* said that "with the dirt removed its true color was revealed"—as white!

47. *The Longest Auto Race*, p. 18; and *Reader's Digest*, January 1963, p. 189.

48. This information is from a Dec. 26, 1995, letter to the author from Ralph Dunwoodie, who supervised the restoration of the car for Harrah's Automobile Collection in Reno, Nev. Dunwoodie said that, as the restoration began, Schuster visited Reno to confirm the car's model number and horsepower rating.

49. Ad by Harry S. Houpt Co., Thomas agent, *New York Times*, Feb. 16, 1908, III, p. 3:1; "Thomas Paris Champion a Stock Chassis," *Automobile*, March 5, 1908, p. 322:1; and "American Car Is All Ready," *New York Times*, Feb. 11, 1908, p. 2:3.

50. But in his Dec. 26, 1995, letter to the author, Dunwoodie indicated that a 35-gallon tank was added to the Thomas, which also carried the standard 22-gallon gas tank.

51. "American Car Here Yesterday," *Cedar Rapids (Iowa) Evening Gazette*, March 2, 1908, n.p.

52. E.M. West, "The New York-Paris Race," *MoTor*, May 1908, p. 40.

53. *New York Times*, "Thomas Car May Enter," Dec. 2, 1907, p. 2:2, and "Montague Roberts in Race," Jan. 3, 1908, p. 1:5.

54. "Montague Roberts Is Dead at 74; Won N.Y.–Paris Auto Race in '08," *New York Times*, Sept. 21, 1957, p. 19:5; and *The Longest Auto Race*, p. 28.

55. "Winning of a 24 Hour Record Race," *Automobile*, Aug. 15, 1907, pp. 221–26.

56. Montague Roberts, "Success in Racing Means Preparation," *New York Times*, Jan. 3, 1909, IV, p. 7:1.

57. *The Longest Auto Race*, pp. 12–15.

58. *New York Times*, Feb. 2, 1908, IV, p. 1:2.

59. "Will Start Alone," *New York Times*, Feb. 9, 1908, II, p. 2:4.

60. *New York Times*, Feb. 9, 1908, IV, p. 1:2. "Giulio" Sirtori appears on several of his *New York Times* dispatches.

61. *Round the World in a Motor-Car*, p. 47.

62. *New York Times*, "Paris Auto Racers Eager for the Start," Feb. 7, 1908, p. 1:1, and Feb. 11, 1908, p. 2:3.

63. "They're Off in Race to Paris," *New York Times*, Feb. 13, 1908, p. 1:7. *Motor Age* estimated the Times Square crowd at "perhaps 10,000."

64. "Vast Throng Sees Start," *New York Times*, Feb. 13, 1908, p. 2:3.

65. "Arctic Dress for the Men," *New York Times*, Feb. 13, 1908, p. 2:7.

66. *New York Times*, Feb. 13, 1908, p. 2:3.

67. *Round the World in a Motor-Car*, p. 28; and *The Longest Auto Race*, p. 22.

68. In various articles on Feb. 9, 1908, II, p. 2, the *New York Times* spells Maurice "Berthe" correctly in one location, but elsewhere refers to "Berkle" and "Berlhe."

69. "Moving Pictures of Start," *New York Times*, Feb. 15, 1908, p. 2:5.

70. "Last Cars Leave Chicago," *New York Times*, March 8, 1908, II, p. 2:5.

71. Schuster in a Jan. 24, 1964, letter to Ralph Dunwoodie, manager of Harrah's Automobile Collection, Reno, Nev.

72. "Autos Fight Snow Drifts," *New York Times*, Feb. 13, 1908, p. 3:3.

73. *New York Times*, Feb. 13, 1908, p. 2:1.

74. "Daring Autoists Ready for Start To-Day for Paris," *New York World*, Feb. 11, 1908, p. 3:1.

75. *New York World*, Feb. 11, 1908, p. 3:1.

76. *Mad Motorists*, p. 66.

77. Unattributed biographical and quoted matter in the foregoing paragraph is from the *New York World*, Feb. 9, 1908, p. 1:4, "Autoists in Dash to Pacific Will Detour to South," Feb. 10, 1908, p. 3:1, and Feb. 11, 1908, p. 3:1.

78. "Vast Throng Sees Autoists Begin 22,000-Mile Journey," *New York World*, Feb. 12, 1908, p. 1:3.

79. "Bad Roads Delay the Werner Car," *New York World*, Feb. 13, 1908, p. 14:5.
80. "Driver Quits Werner Car," *New York Times*, Feb. 16, 1908, II, p. 3:4.
81. *New York Times*: Untitled wire report, Feb. 22, 1908, p. 2:6, "Evil Thoughts of Sirtori," Feb. 16, 1908, II, p. 3:1, "The Protos at Albany," Feb. 14, 1908, p. 2:2, and "St. Chaffray's Story," Feb. 14, 1908, p. 2:1.
82. C.A Briggs, "The Road to Paris," cartoon, *Chicago Daily Tribune*, Feb. 27, 1908, p. 6:2.
83. "Thomas First at Buffalo," *New York Times*, Feb. 17, 1908, p. 2:1.
84. "Cars All Right—Roberts," *New York Times*, Feb. 16, 1908, II, p. 3:2.
85. T.R. Nicholson, *Adventurer's Road: The Story of Pekin–Paris, 1907, and New York–Paris, 1908* (New York: Rinehart & Co., 1958), p. 147.
86. "Thomas Starts Again," *New York Times*, Feb. 21, 1908, p. 2:3.
87. "German Car Lightened," *New York Times*, Feb. 16, 1908, II, p. 2:6.
88. *New York Times*, Feb. 17, 1908, p. 2:1, and "Zust Car Has Mishap," Feb. 17, 1908, p. 2:4.
89. "Two Cars Reach Erie," *New York Times*, Feb. 18, 1908, p. 2:1.
90. *New York Times*, Feb. 13, 1908, p. 3:2.
91. "Two Autos Reach Fonda," *New York Times*, Feb. 14, 1908, p. 2:1.
92. *New York Times*, Feb. 16, 1908, II, p. 3:2.
93. *New York Times* editorials, "Having Fun With the Automobilists," Feb. 15, 1908, p. 6:3, and "Disgraced by Our Bad Roads," Feb. 17, 1908, p. 6:4.
94. *New York Times*, "Three Autos at Canastota," Feb. 15, 1908, p. 1:7, and "St. Chaffray's Popularity," Feb. 15, 1908, p. 2:3.
95. *New York Times*, Feb. 15, 1908, p. 1:7; and *The Longest Auto Race*, p. 29.
96. "The Men and Cars," *New York Times*, Feb. 16, 1908, II, p. 2:3.
97. *Round the World in a Motor-Car*, p. 34.
98. *New York Times*, Feb. 16, 1908, II, p. 2:2.
99. These biographical details are from the *New York Times*, "T. Walter Williams of the Times Dies," Nov. 10, 1942, p. 27:1, and Percy S. Bullen, "A Grand Man Has Left Us," Nov. 12, 1942, p. 24:7.
100. "Leaders Stuck in Swamp," *New York Times*, Feb. 16, 1908, II, p. 1:7.
101. *New York Times*, Feb. 16, 1908, II, p. 2:2. In his book, however, Schuster recalls that four horses pulled the Thomas "from the mud near Seneca Falls" (p. 30)—evidently later that same day.
102. *New York Times*, "St. Chaffray's Own Story," Feb. 15, 1908, p. 2:2, and Feb. 16, 1908, II, p. 2:3.
103. *New York Times*, Feb. 17, 1908, p. 1:7.
104. "Zust Catches De Dion," *New York Times*, Feb. 22, 1908, p. 2:5.
105. "No Rest for Zust Crew," *New York Times*, Feb. 19, 1908, p. 2:4.
106. Rushmore headlamp ad, *Automobile*, Dec. 19, 1907, p. 95:1.
107. *New York Times*, Feb. 19, 1908, p. 2:4.
108. *New York Times*, Feb. 17, 1908, p. 2:1.
109. *New York Times*, Feb. 18, 1908, pp. 1:7, 2:1.
110. "From the Man in the Lead," *New York Times*, March 15, 1908, IV, p. 1:2; and *The Longest Auto Race*, p. 32.

111. *The Longest Auto Race*, p. 33; and "Motor Race Jars Peace of Nations," *Chicago Daily Tribune*, March 2, 1908, p. 3:1. By "hippodrome race," Roberts was referring to a contest with a prearranged winner.

112. *Cedar Rapids (Iowa) Evening Gazette*, March 2, 1908, n.p.

113. "Albany Cheers Moto-Bloc," *New York Times*, Feb. 15, 1908, p. 2:4.

114. "Sizaire-Naudin Fixed," *New York Times*, Feb. 15, 1908, p. 2:4.

115. *New York Times*, "Sizaire-Naudin on the Way," Feb. 17, 1908, p. 2:5, and "Sizaire-Naudin Withdrawn," Feb. 19, 1908, p. 2:5.

116. "Drivers Hurry Onward," *New York Times*, Feb. 19, 1908, p. 2:2.

117. *New York Times*, Feb. 19, 1908, pp. 2:2, 2:3.

118. *New York Times*, "Hunger Stops St. Chaffray," Feb. 19, 1908, p. 2:3, and "St. Chaffray Cheerful," Feb. 17, 1908, p. 2:3.

119. "Thomas Car Now 51 Miles Ahead," *New York Times*, Feb. 23, 1908, II, p. 2:3.

120. *New York Times*, Feb. 19, 1908, p. 2:4.

121. "Driving through Blizzard," *New York Times*, Feb. 20, 1908, p. 1:7.

122. *New York Times*, Feb. 21, 1908, pp. 1:7, 2:3.

123. "Thomas Car Leads 26 Miles in Race," *New York Times*, Feb. 22, 1908, p. 2:4. To identify the hometown of Buggins's Shoe Emporium, I wrote to public libraries in the Indiana towns along the Thomas route. None was able to find a reference to the store. According to the Indiana State Library, Rand McNally's 1907 *Photo-Auto Guide Book* contains a photograph of every turn along the route; none shows a Buggins's Shoe Emporium sign. Was "Buggins's" a misprint? The 1906 Elkhart, Ind., city directory lists a shoemaker named Edgar H. Huggins. The 1908 Elkhart city directory, however, lists neither Huggins nor a Huggins's Shoe Emporium.

124. *New York Times*, Feb. 23, 1908, II, p. 1:1.

125. "Officials Were Absent," *New York Times*, Feb. 25, 1908, p. 2:4.

126. *New York Times*, Feb. 23, 1908, II, p. 2:2.

127. "Italian Crew Presses On," *New York Times*, Feb. 23, 1908, II, p. 2:3.

128. "8 Miles in 22 Hours," *New York Times*, Feb. 25, 1908, p. 1:1.

129. "Statement of E.R. Thomas," *New York Times*, March 3, 1908, p. 2:5.

130. "Chicago Triumph for Leading Auto," *Chicago Daily Tribune*, Feb. 26, 1908, p. 3:1.

131. *Chicago Daily Tribune*, Feb. 26, 1908, p. 3:1.

132. *Chicago Daily Tribune*, Feb. 26, 1908, p. 3:1; and "Chicago Welcomes First Paris Racer," *New York Times*, Feb. 26, 1908, p. 1:1.

133. *Chicago Daily Tribune*, Feb. 26, 1908, p. 3:1; and "Thomas Car's Journey Faster," *New York Times*, Feb. 23, 1908, II, p. 2:6.

134. "St. Chaffray Is Happy," *New York Times*, Feb. 27, 1908, p. 2:4.

135. "Motobloc Closing Up," *New York Times*, Feb. 24, 1908, p. 2:5.

136. *New York Times*, "Protos at Cleveland," Feb. 21, 1908, p. 2:4, and "German Crew Is Weary," Feb. 26, 1908, p. 2:6.

137. "Protos Car Delayed," *New York Times*, Feb. 28, 1908, p. 2:3.

138. *The Longest Auto Race*, p. 39. The housings had several functions, Schuster explained (p. 54): "These two half-bell-shaped castings were riveted to the outside of the chassis

frame, they carried the shackled front end of the rear springs, also the front end of the radius rod which kept our driving chains in adjustment. Also the outer end of the transmission countershaft carrying the driving sprockets had their bearings in these housings.... To install new housings, it was necessary to take the body from the chassis before they could be riveted to the frame sides."

139. "Thomas to Paris Even If No Reward," *New York Times*, March 1, 1908, II, p. 2:2.
140. All five charges are from the *New York Times*, March 1, 1908, II, p. 2:2.
141. *New York Times*, March 15, 1908, IV, p. 1:3.
142. "Statement by E.R. Thomas," *New York Times*, March 1, 1908, II, p. 2:3.
143. *New York Times*, "Montague Roberts Replies," March 1, 1908, II, p. 2:3, and "What the Thomas Car Did," March 2, 1908, p. 2:6.
144. *Chicago Daily Tribune*, March 2, 1908, p. 3:1.
145. *Chicago Daily Tribune*, March 2, 1908, p. 3:1, and "World Racers Lost in City," March 4, 1908, p. 3:4.
146. *New York Times*, March 2, 1908, p. 2:6.
147. *New York Times*, March 1, 1908, II, p. 2:2.
148. "Arctic Explorer Quits French Car," *Chicago Daily Tribune*, Feb. 28, 1908, p. 4:5.
149. "The Men Who Drive the American Car," *Omaha (Neb.) World-Herald*, March 8, 1908, p. 6M:6.
150. "St. Chaffray Says Hansen Is Imposter," *Omaha (Neb.) Daily News*, March 10, 1908, p. 6:1. St. Chaffray is referring to Fridtjof Nansen (1861–1930) of Norway, a pioneer oceanographer, zoologist, statesman and polar explorer. His name lives on in the Nansen bottle (for capturing seawater at a given depth), Nansen stove (one of many items the *New York Times* recommended that the racers carry), and Nansen sledge.
151. Dermot Cole, *Hard Driving: The 1908 Auto Race from New York to Paris* (New York: Paragon House, 1991), p. 71.
152. His name appeared as "Schneider" as often as "Snyder." I've chosen the Snyder spelling for two reasons: One, Koeppen spells his teammate's name as Snyder. Two, the 1908 Chicago city directory lists an Oscar W. Snyder, "auto operator," but contains no Schneider entry that corresponds with the initials O.W.
153. "Will Leave Chicago To-Day," *New York Times*, March 7, 1908, p. 2:4; and "German Car Stalled in Barn," *Chicago Daily Tribune*, March 8, 1908, p. L5:5.
154. "German Interest in Race," *New York Times*, March 9, 1908, p. 2:3.
155. *New York Times*, "Koeppen's Fortune Used in Auto Race," April 16, 1908, p. 6:1, untitled article, March 10, 1908, p. 2:4, and April 16, 1908, p. 6:1.
156. "'Hulfe!' Also 'Au Secours!'/Macedonian Cry for Help from Foreign Autoists in Indiana," *Chicago Daily Tribune*, March 1, 1908, p. 1:4.
157. *New York Times*, "Motobloc Reaches Chicago," March 4, 1908, p. 2:4, and "Protos Reaches Chicago," March 5, 1908, p. 2:5.
158. Untitled article, *New York Times*, March 2, 1908, p. 2:5; and *Chicago Daily Tribune*, March 1, 1908, p. 1:4.
159. "Patriotism Far Astray," editorial, *New York Times*, March 1, 1908, II, p. 8:3.
160. "Motobloc at South Bend," *New York Times*, Feb. 28, 1908, p. 2:3; and *Chicago Daily Tribune*, March 4, 1908, p. 3:4.

161. *New York Times*, "Zust Races De Dion to Chicago Club," Feb. 27, 1908, p. 2:4, and Feb. 26, 1908, p. 2:5.
162. "To Find Motobloc Thieves," *New York Times*, Feb. 28, 1908, p. 2:3.
163. "Moto Bloc's Stolen Articles Found," *Chicago Daily Tribune*, March 8, 1908, p. L5:5.
164. *New York Times*, March 1, 1908, II, p. 8:3.
165. *New York Times*, "St. Chaffray's Story," Feb. 21, 1908, p. 2:3, "Auto Racer across Mississippi River," March 1, 1908, II, p. 2:4, and untitled article, Feb. 19, 1908, p. 2:4.
166. *New York Times*, March 1, 1908, II, p. 1:1; and "American Car Spends Night at Logan, Ia.," *Omaha (Neb.) World-Herald*, March 4, 1908, p. 1:5.
167. *The Longest Auto Race*, p. 43.
168. "Autos Struggle on through Iowa Mud," *New York Times*, March 2, 1908, p. 1:1.
169. *New York Times*, March 2, 1908, p. 1:1.
170. "Auto Race Leader Close to Omaha," *New York Times*, March 4, 1908, p. 1:1.
171. *Omaha (Neb.) World-Herald*, March 8, 1908, p. 6M:2.
172. "Omaha Crowds Meet Auto Race Leader," *New York Times*, March 5, 1908, p. 1:1.
173. *New York Times*, March 5, 1908, p. 1:1.
174. *Omaha (Neb.) World-Herald*, March 5, 1908, p. 5:3; and "Yankee Car Resumes Trip," *Omaha (Neb.) Daily Bee*, March 6, 1908, p. 9:3.
175. *Omaha (Neb.) Daily Bee*, March 6, 1908, p. 9:3.
176. "Nebraska Mud Now for Auto Racers," *New York Times*, March 6, 1908, p. 1:1.
177. "Impressions of a Pilot," *New York Times*, March 8, 1908, II, p. 2:5.
178. *New York Times*, March 15, 1908, IV, p. 1:3.
179. "Auto Race Leader Makes 140 Miles," *New York Times*, March 7, 1908, p. 1:1.
180. *New York Times*, March 8, 1908, II, p. 2:6.
181. "Cowboys Welcome Racer at Cheyenne," *New York Times*, March 9, 1908, p. 1:1.
182. *New York Times*, March 5, 1908, p. 1:1.
183. "Order of the Cars at Various Laps," *New York Times*, April 25, 1908, p. 5:2; and "Strang Wins Briarcliff Trophy Race," *Cycle and Automobile Trade Journal*, May 1908, pp. 25–27.
184. "New Driver of the Thomas," *New York Times*, March 9, 1908, p. 2:4; and "Fastest Express Train Outsped By Motor Car," *Rocky Mountain News* (Denver), March 23, 1907, p. 8:1.
185. "American Car Delayed by a One-Man Strike," *Cheyenne (Wyo.) Daily Leader*, March 10, 1908, p. 1:1.
186. "Auto Racer Leader in Wyoming Snow," *New York Times*, March 11, 1908, p. 1:1.
187. "Mountain Scaled by Leading Auto," *New York Times*, March 13, 1908, p. 1:1.
188. "Current Comment," *Motor Age*, March 12, 1908, p. 9:3.
189. *The Longest Auto Race*, pp. 57-59.
190. "Utah Snows Halt the Leading Auto," *New York Times*, March 15, 1908, II, p. 1:1.
191. "Denver Race Is Won by Thomas-Detroit," *Motor Age*, June 4, 1908, p. 20. Details were also drawn from "10,000 See Mathewson Win Big Auto Race," *Denver Post*, May 31, 1908, pp. 1–2, which identifies the car as a Thomas 40. Other sources suggest Mathewson's car may have actually been a Thomas-Detroit or a Chalmers-Detroit.

192. *New York Times*, March 9, 1908, p. 2:4.

193. "Galveston Beach Races," *Automobile*, Aug. 11, 1910, pp. 252-53; and "Cadillac Sets Another Record," *Detroit News Tribune*, July 16, 1916, p. 9:8 of automobile section. Other details on Brinker's background are from the *New York Times*, Feb. 11, 1908, p. 2:4; "Smashed Denver Record," *Automobile*, Feb. 11, 1909, p. 298:1; *Wyoming Tribune* (Cheyenne), Aug. 13, 1910, "National Racer Here" and "Harold Brinker Returns from Racing in Galveston," both p. 2:3; "Birdmen at Cheyenne," *Rock Springs (Wyo.) Rocket*, July 28, 1911, p. 1:3; and "Harold Brinker, Early Denver Auto Racer, Dies at 69," *Rocky Mountain News* (Denver), May 8, 1955, p. 23.

194. *The Longest Auto Race*, p. 73.

195. "Auto Racer Nears Great Gold Camp," *New York Times*, March 19, 1908, p. 1:1.

196. Untitled article, *New York Times*, March 24, 1908, p. 1:1.

197. "San Francisco Wild over Race Leader," *New York Times*, March 25, 1908, p. 1:1.

198. *New York Times*, March 15, 1908, IV, p. 1:1.

199. "Two Racers at Rochelle," *New York Times*, March 1, 1908, II, p. 2:4.

200. *Round the World in a Motor-Car*, p. 117.

201. "World's Racers Pierce the West," *Motor Age*, March 12, 1908, p. 13:1.

202. *Round the World in a Motor-Car*, pp. 80–81.

203. *Round the World in a Motor-Car*, pp. 80, 82.

204. "Zust Travels on Railway," *New York Times*, March 7, 1908, p. 1:1.

205. "Italian Car Arrives After Hard Struggle," *Omaha (Neb.) World-Herald*, March 8, 1908, p. 8N:5.

206. "Experiences of the Italian Autoists," *Omaha (Neb.) Daily News*, March 8, 1908, p. C5:3.

207. "Italian Car Leaves Omaha on its Long Trip to Paris," *Omaha (Neb.) Daily News*, March 9, 1908, p. 7:1; and "Zust Gains on Leader," *New York Times*, March 10, 1908, p. 1:1.

208. *Round the World in a Motor-Car*, p. 87.

209. *Round the World in a Motor-Car*, pp. 89, 90; and "Zust Gains on the Leader," *New York Times*, March 14, 1908, p. 1:1.

210. "Italians Have Difficulty," *New York Times*, March 15, 1908, II, p. 2:3.

211. "Zust's Narrow Escape," *New York Times*, March 19, 1908, p. 1:1; and Round the World in a Motor-Car, p. 109.

212. "Crew Near Death on the Zust Car/Four Times the Italian Auto in Race to Paris Has Narrow Escapes," *New York Times*, March 29, 1908, IV, p. 3:1.

213. *Round the World in a Motor-Car*, p. 104.

214. *Round the World in a Motor-Car*, p. 112.

215. *New York Times*, March 29, 1908, IV, p. 3:1.

216. "Union Pacific Explains," *New York Times*, March 22, 1908, II, p. 2:1.

217. "Italians on Zust Kill 20 Wolves," *New York Times*, March 21, 1908, p. 1:1; and *Round the World in a Motor-Car*, pp. 147–48.

218. *Round the World in a Motor-Car*, p. 117.

219. *Round the World in a Motor-Car*, pp. 121, 124; and "Railroads Aid the Zust," *New York Times*, March 25, 1908, p. 2:5.

220. *New York Times*, March 29, 1908, IV, p. 3:1.
221. *Round the World in a Motor-Car*, pp. 142–44.
222. "Los Angeles Wild over the Zust Car," *New York Times*, April 1, 1908, p. 2:4.
223. "Second Auto Racer at San Francisco," *New York Times*, April 5, 1908, II, p. 18:2. The *Times* also spelled it "Vollmueller."
224. *Round the World in a Motor-Car*, pp. 142–44.
225. Scarfoglio gives the 4,600-mile figure in "California Like Italy," *New York Times*, April 8, 1908, p. 2:4. In his book (p. 150), he inflates the mileage to an even 5,000.
226. "De Dion Breaks Shaft," *New York Times*, March 10, 1908, p. 2:2.
227. *New York Times*, "De Dion's Fight with Mud," March 14, 1908, p. 2:4, and "De Dion's Plucky Run," March 13, 1908, p. 2:6.
228. *New York Times*, March 13, 1908, p. 2:6.
229. "French Car Laid Up for Day or Two," *Omaha (Neb.) World-Herald*, March 14, 1908, p. 1:6; and "St. Chaffray Fixing Broken Pinion," *Omaha (Neb.) Daily News*, March 14, 1908, p. 7:1.
230. *Omaha (Neb.) World-Herald*, March 14, 1908, p. 1:6; and "De Dion Crossing Nebraska," *New York Times*, March 17, 1908, p. 2:4.
231. "St. Chaffray Is Some Spender," *Omaha (Neb.) Daily News*, March 13, 1908, p. 14:3.
232. *New York Times*, untitled article, March 20, 1908, p. 2:5, and "De Dion at Cheyenne," March 21, 1908, p. 2:3.
233. "De Dion Escapes Trains," *New York Times*, March 26, 1908, p. 2:3.
234. Three times in his daily *New York Times* dispatches (March 27, March 28 and April 5), St. Chaffray writes of this adventure. He places it, first, east of Evanston; second, "after leaving Evanston" (west of the town); and, third, at Spring Valley.
235. *New York Times*, untitled article, March 27, 1908, p. 2:5, and "De Dion Leaves Ogden," March 28, 1908, p. 2:4.
236. "St. Chaffray in the Desert," *New York Times*, March 30, 1908, p. 5:3.
237. "De Dion at Goldfield," *New York Times*, April 1, 1908, p. 2:6.
238. St. Chaffray relates these events in "French Car in Peril in Death Valley," *New York Times*, April 4, 1908, p. 1:1. Other information was drawn from a news story, "De Dion 22 Hours in Death Valley," *New York Times*, April 3, 1908, p. 2:3.
239. Untitled article, *New York Times*, April 5, 1908, II, p. 18:5.
240. "Third Auto Racer at San Francisco," *New York Times*, April 8, 1908, p. 2:3.
241. "A Frenchman's View of the New York–Paris Race," *Automobile Topics*, April 18, 1908, pp. 114-15.
242. Untitled, *Automobile Topics*, May 9, 1908, p. 293.
243. "Motobloc Loses Its Way," *New York Times*, March 16, 1908, p. 2:4.
244. "Motobloc at Ogden, Iowa," *New York Times*, March 17, 1908, p. 2:5.
245. "Autos Forbidden to Go by Train," *Chicago Daily Tribune*, March 5, 1908, p. 9:1.
246. "Ship French Auto Ahead by Freight," *Omaha (Neb.) World-Herald*, March 19, 1908, p. 11:3.
247. "Motobloc Sent on by Train," *New York Times*, March 19, 1908, p. 2:6; and *Omaha (Neb.) World-Herald*, March 19, 1908, p. 11:3.

248. "Protos Sheds Weight," *New York Times*, March 12, 1908, p. 2:2.
249. "Protos Waits for Tires," *New York Times*, March 15, 1908, II, p. 2:6; and "French Car Still in Iowa," *Omaha (Neb.) Daily Bee*, March 14, 1908, p. 2:5.
250. Quoted in "Lieutenant Koeppen Tells the Story of the New York-to-Paris Race," *New York Times*, Nov. 22, 1908, V, p. 5:4.
251. "German Car Arrives 'Midst Pandemonium," *Omaha (Neb.) World-Herald*, March 18, 1908, p. 4:3; and "German Car in Omaha," *New York Times*, March 18, 1908, p. 2:3.
252. "Germans Reach Cheyenne," *New York Times*, March 22, 1908, II, p. 2:4; and *The Longest Auto Race*, p. 50.
253. "Last of the Racing Cars," *Ogden (Utah) Standard*, April 3, 1908, p. 8:2.
254. "Troubles of German Crew," *New York Times*, April 8, 1908, p. 2:4. The Central Pacific became the Southern Pacific. Either name applies to the route between Kelton and Ogden over which the Union Pacific sent out special trains to rescue first the Zust and later the Protos.
255. "New York Paris Cars Will Sail from Seattle to Vladivostok—Race Will Be Resumed," *Motor Age*, April 16, 1908, p. 16:1.
256. "German Car on Flat Car," *Pocatello (Idaho) Tribune*, April 13, 1908, p. 1:4.
257. One wire report suggests that Snyder drove the Protos ahead to Pocatello while Koeppen waited in Seattle. But in his book, *Im Auto um die Welt*, pp. 153–54, Koeppen makes it clear that the car was shipped from Ogden. He asserts, however, that instead of being repaired in Ogden, it was shipped from there "in the same condition in which it had been brought from the site of our last accident beyond Kelton."
258. As described in "German Auto First in Paris," *New York Times*, July 27, 1908, p. 1:7.
259. "New York–Paris Racer Sold at Auction," *Automobile Topics*, May 2, 1908, p. 251.
260. "Zust Car in Paris," *New York Times*, Sept. 18, 1908, p. 1:4.
261. "Pathfinder Car Ends Trips from Ocean to Ocean," *Seattle Post-Intelligencer*, May 20, 1909, p. 2:3; and "New York-to-Paris Winner Still Going," *Automobile*, April 25, 1912, p. 1003:2.
262. *The Longest Auto Race*, p. 148.
263. Much of this chronology is from a March 13, 1964, press release from Harrah's.
264. According to *The Longest Auto Race*, p. 147, and *Hard Driving*, p. 220.
265. *Reader's Digest*, January 1963, p. 188:1.
266. *Hard Driving*, p. 21.
267. *The Longest Auto Race*, p. 143.
268. *The Longest Auto Race*, p. 143. Schuster refers to *The Story of the New York to Paris Race* (Buffalo, N.Y.: E.R. Thomas Motor Co., 1908; reprinted Los Angeles, Calif.: Floyd Clymer, 1951), 75 pp. Besides those mentioned elsewhere in this chapter, other books about the New York–Paris race include Alise Barton Whiticar, *The Long Road: The Story of the Race around the World by Automobile in 1908* (Port Salerno, Fla.: Whiticar Books, 1973), 355 pp., and Robert B. Jackson, *Road Race Round the World: New York to Paris, 1908*, 2nd ed. (New York: Henry Z. Walck Inc., 1977), 52 pp.
269. "The Automobile Race from New York to Paris," *Current Literature*, April 1908, pp. 359, 361.

270. "The New York to Paris Race," *Outing Magazine*, October 1908, p. 118:2.

271. "The New York to Paris Race Around the World," *Overland Monthly*, April 1908, p. 370:2.

272. *New York Times*, March 4, 1908, p. 1:1; and "Big Auto Race Puts Life into Auto Race," *Omaha (Neb.) World-Herald*, March 16, 1908, p. 2:3.

273. As quoted in "Moral from Auto Contest," *New York Times*, Feb. 17, 1908, p. 2:5.

274. Spicer ad, *SAE Journal*, February 1952, p. 114; and Siemens ad, *Fortune*, April 18, 1994, pp. 240–41.

275. T. Walter Williams in James Rood Doolittle, ed., *The Romance of the Automobile Industry* (New York: Klebold Press, 1916), p. 401.

276. The Lincoln Highway Association, *The Lincoln Highway: The Story of a Crusade That Made Transportation History* (New York: Dodd, Mead & Co., 1935), p. 113.

277. "Autoists Eager for Race," *New York Times*, Nov. 29, 1907, p. 1:1.

278. "Auto Race a Study in Public Schools," *New York Times*, March 17, 1908, p. 2:6.

279. "A Race and More," editorial, *New York Times*, May 13, 1908, p. 6:3.

280. *New York Times*, "Great Auto Race Planned for August," May 16, 1908, p. 1:3, "English Autoists Praise Big Tour," June 7, 1908, IV, p. 4:4, and "Roberts May Enter New Race," May 20, 1908, p. 1:1.

281. "M.C.A. Won't Support Cross-Country Test," *Automobile*, April 1, 1909, p. 551:1.

Notes to Chapter 3

1. H.B. Harper, *The Story of the Race* (Detroit: Ford Motor Company, 1909?), p. 6. (According to the title page of the booklet, its unidentified author was "one of the Crew on Ford Car No. 1." The vote, then, is between Frank Kulick, the race driver, and H.B. Harper, the Ford Motor Company's advertising manager. Harper wins.)

2. *Story of the Race*, p. 6. M.M. Musselman summarizes the history of the Selden Patent and its fallout in "The Villain of Gasoline Alley," Chapter 2 of *Get a Horse! The Story of the Automobile in America* (Philadelphia: J.B. Lippincott Co., 1950), pp. 24–32.

3. "Tell Harmon of Long Motor Run," *Cleveland Leader*, May 2, 1909, III, p. 4:1.

4. "Scott First to Reach Seattle," *Motor World*, June 24, 1909, p. 500:2.

5. Biographical information is from "Financier Heir to Guggenheim Fortune Dies," *Rocky Mountain News* (Denver), Nov. 18, 1959, n.p.; and *Who Was Who in America* (Chicago: Marquis-Who's Who Inc., 1963), pp. 352–53. Born May 17, 1885, Guggenheim was just 23 when the race was announced, 24 by the start.

6. "For Continental Auto Contest Every Summer," *Seattle Post-Intelligencer* June 30, 1909, p. 2:3.

7. "Cross Country Run Planned," *Motor Age*, Feb. 18, 1909, p. 18:2.

8. "New York to Seattle Program," *Motor World*, March 8, 1909, p. 1145:1; and "New York-Seattle Run," *Motor Age*, March 4, 1909, p. 4:3.

9. "Automobilists Want Transcontinental Highway," *New York Times*, April 18, 1909, IV, p. 4:7, reports 116 entries in the design contest. But "Silversmith Has Auto Cup Ready," *Seattle Post-Intelligencer*, June 20, 1909, p. 8:2, refers to 20 entries. Most press accounts valued the trophy at $2,000. Exceptions are the *New York Times*,

which says $2,250, and the Ford Motor Company, which in *Story of the Race*, p. 2, says $3,500. The trophy "was encrusted with nuggets of Yukon gold and the precious metals in it were alone worth $2,000," the automaker claimed 50 years later in its *Chronological Story of the New York-to-Seattle Transcontinental Endurance Race, June 1–June 23, 1909* (Dearborn, Mich.: Ford Motor Co., May 1959), p. 1. "Mr. Guggenheim paid another $1,500 to Shreve & Company for designing and building it. Thus, it was a '$3,500 trophy.'"

10. *New York Times*, April 18, 1909, IV, p. 4:7.
11. *New York Times*, April 18, 1909, IV, p. 4:7.
12. Benjamin Briscoe, "Relation of M.C.A. to Auto Industry," *New York Times*, April 11, 1909, IV, p. 4:1.
13. Twenty-seven U.S. automakers voted against the New York–Seattle race, according to "M.C.A. Won't Support Cross-Country Test," *Automobile*, April 1, 1909, p. 551:2. Of these 27, all but three belonged to the Association of Licensed Automobile Manufacturers as of Jan. 15, 1910, according to "A.L.A.M. Now Licenses Sixty-Two Makes," *Automobile*, Jan. 20, 1910, p. 171:1. Two of the three non-members were out of business by 1910.
14. "Makers Refuse Aid to Seattle Race," *New York Times*, March 31, 1909, p. 12:4.
15. *New York Times*, March 31, 1909, p. 12:4; *Cleveland Leader*, May 2, 1909, III, p. 4:1; and "Stearns Opposes Seattle Contest," *New York Times*, April 22, 1909, p. 10:3.
16. "Simplex for Seattle," *New York Times*, April 3, 1909, p. 7:6.
17. *New York Times*, April 22, 1909, p. 10:3.
18. *New York Times*, April 22, 1909, p. 10:3.
19. "Shouldn't Be a Race of Dollars," editorial, *New York Times*, April 23, 1909, p. 8:4.
20 "Ford Favors Ocean to Ocean Contest," *New York Times*, April 12, 1909, p. 8:5.
21. "Will Show Need of Good Roads," *Seattle Post-Intelligencer*, June 6, 1909, p. 3:4 of sports section.
22. "Ford Coast-to-Coast Car," *Motor Age*, Oct. 7, 1909, p. 16:2.
23. "That Ocean to Ocean Contest," *Ford Times*, May 15, 1909, pp. 6–7.
24. *Chronological Story of the New York-to-Seattle Transcontinental Endurance Race*, p. 8.
25. The *Transcontinental Automobile Guide*, p. 22, joins a number of contemporary press accounts in spelling it "Reddington." Confusion reigned, however, for other sources said "Redington." The *New York Times* and *Seattle Post-Intelligencer* only added to the confusion by spelling it both ways.
26. "L.W. Redington to Talk on Good Roads," *Seattle Post-Intelligencer*, May 28, 1909, p. 12:3.
27. "Financier Heir to Guggenheim Fortune Dies," *Rocky Mountain News* (Denver), Nov. 18, 1959, n.p.; and *New York Times*, April 18, 1909, IV, p. 4:1.
28. "Coast-to-Coast Route Set," *Motor Age*, March 25, 1909, p. 10:3.
29. So identified in *Transcontinental Automobile Guide*, p. 22, and several press accounts, though other sources spelled it "Ely."
30. "Pathfinder at Work," *Motor Age*, March 25, 1909, p. 10:3.
31. "Pathfinder In Chicago," *Motor Age*, April 1, 1909, p. 15:3; and "Pathfinder Off Again Today," *Chicago Daily Tribune*, April 1, 1909, p. 14:5.

32. *New York Times*, April 18, 1909, IV, p. 4:1; and "Seattle Pathfinder Reaches Colorado," *Automobile*, April 15, 1909, p. 608:2.

33. *Motor Age*, March 25, 1909, p. 10:3.

34. *New York Times*, April 18, 1909, IV, p. 4:3, which gives Washington state's appropriation as $125,000; *Motor Age* said $120,000.

35. "National Road Proposed," *Motor Age*, May 6, 1909, p. 11:3; and "Cross the Country Mud-Plugs the Thomas," *Automobile*, April 8, 1909, p. 592:2.

36. "New Phase in the New York to Seattle Contest," *Automobile*, April 22, 1909, p. 647:1; and "Want Road from Coast to Coast," *Seattle Post-Intelligencer*, May 21, 1909, p. 12:2.

37. "Thomas Pathfinder Reaches Idaho," *Automobile*, April 29, 1909, p. 713:2.

38. "Pathfinder Car Ends Trip from Ocean to Ocean," *Seattle Post-Intelligencer*, May 20, 1909, p. 2:3.

39. *Seattle Post-Intelligencer*, May 20, 1909, p. 1:7.

40. *Automobile*, April 29, 1909, p. 713:2.

41. "Pathfinder In Trouble," *Motor Age*, May 6, 1909, p. 11:1.

42. *Seattle Post-Intelligencer*, May 20, 1909, p. 2:3.

43. George N. Schuster and Tom Mahoney, *The Longest Auto Race* (New York: John Day Co., 1966), p. 142.

44. *Seattle Post-Intelligencer*, May 20, 1909, p. 1:7.

45. *Seattle Post-Intelligencer*, May 20, 1909, p. 2:4.

46. *Seattle Post-Intelligencer*, May 20, 1909, p. 2:4.

47. "Perils of Pathfinding in West," *New York Times*, May 23, 1909, IV, p. 4:3.

48. *New York Times*, April 18, 1909, IV, p. 4:2.

49. Herbert C. Lanks, *Highway to Alaska* (New York: D. Appleton-Century Co., 1944), p. 24.

50. *Chronological Story*, p. 3.

51. "Sanction for Auto Race," *New York Times*, March 1, 1909, p. 7:2.

52. "Five Cars Start in Seattle Contest," *New York Times*, June 2, 1909, p. 3:5. What happened to the eight official entries that failed to start on June 1? The Simplex Automobile Company withdrew its car, as did seven private owners who had entered two Thomas autos, a Franklin, Garford, Renault, Stearns and Welch. The five cars that started June 1 included the Itala, which entered after the May 15 deadline and thus was not counted among the 13 official entrants. The Stearns owned by Oscar Stolp was the sixth car in the race but started from New York four days behind the other contestants.

53. "But Five Cars Start for Seattle," *Motor World*, June 3, 1909, p. 374:1.

54. *Story of the Race*, p. 7; and "Centenarian Would Be Carnival Queen," *New York Times*, April 16, 1909, p. 7:6.

55. "Five Cars Start in Seattle Contest," *New York Herald*, June 2, 1909, p. 11:3.

56. "Quintette of Cars Leave New York for Seattle," *Automobile*, June 3, 1909, p. 895:1; and Allan Nevins, *Ford: The Times, the Man, the Company* (New York: Charles Scribner's Sons, 1954), pp. 416-17.

57. *New York Herald*, June 2, 1909, p. 11:3.

58. *New York Herald*, June 2, 1909, p. 11:3; "Five Cars Start in New York-Seattle Contest," *Automobile Topics*, June 5, 1909, p. 588:1; "Ocean to Ocean Auto Racers on Way to

Pacific," *Seattle Post-Intelligencer*, June 2, 1909, p. 1:7; and *Automobile*, June 3, 1909, p. 895:1.

59. *New York Times*, June 2, 1909, p. 3:4.

60. *New York Herald*, June 2, 1909, p. 11:4.

61. *Automobile*, June 3, 1909, p. 895:1.

62. "Halt Here on Run to Pacific," *Syracuse (N.Y.) Post-Standard*, June 3, 1909, p. 7:1.

63. "Shawmut Finishes Long Race," *Chicago Daily Tribune*, June 25, 1909, p. 12:6; *Chronological Story*, p. 8; and "Solars on Ford Cars," *Motor Field*, July 1909, p. 55:2.

64. Henry Ford with Samuel Crowther, *My Life and Work* (Garden City, N.Y.: Doubleday, Page & Co., 1923), p. 14.

65. "Tires Hold Well in Long Contest," *Seattle Post-Intelligencer*, June 20, 1909, p. 5:3 of sports section.

66. "New York to Seattle Cars Start Racing," *New York Times*, June 8, 1909, p. 8:5.

67. *Chronological Story*, p. 4; and "Model T Cars in the Ocean to Ocean Contest," *Ford Times*, April 15, 1909, p. 2:2.

68. *Automobile*, July 4, 1907, "That Recent 24-Hour at Detroit," p. 30:1, National ad, p. 89:1, and Ford ad, p. 92:1.

69. Jack B. Scott, *The Big Race* (Raytown, Mo.: Jack B. Scott, 1992), 104 pp., spells it "Burt" Scott. In an Oct. 18, 1995, telephone interview, Jack Scott, the son of Bert Scott of Ford No. 2, told the author that he used the spelling on the Social Security card of his father, who had been dead for many years before the writing of *The Big Race*. Because all other contemporary and retrospective press accounts, as well as Ford Motor Company records and publications, identify him as Bert Scott, this book adopts that spelling.

70. *The Reminiscences of C.J. Smith* (Dearborn, Mich.: Ford Motor Co. Archives, Oral History Section, November 1951), based on a March 23, 1951, interview with C.J. Smith at his home in Dearborn, p. 9.

71. Press accounts don't identify which of its three 1909 6-cylinder models the automaker entered. But according to "Details of the 1909 Cars," *Automobile*, Dec. 31, 1908, just two Acmes had the 48-horsepower engine used in the New York–Seattle racer. Model 20 was a touring car, Model 21 a four-passenger runabout.

72. *Automobile*, June 3, 1909, p. 895:1.

73. *New York Times*, April 18, 1909, IV, p. 4:6.

74. *New York Times*, April 18, 1909, IV, p. 4:6, and "Locomobile Wins Philadelphia Race," Oct. 11, 1908, IV, p. 2:5.

75. *Automobile*, June 3, 1909, p. 895:2; and "The Guggenheim Itala Car Carries Continentals," *Motor Field*, July 1909, p. 118:2.

76. "Tough Traveling to Seattle," *Motor World*, June 17, 1909, p. 457:3.

77. *Automobile*, June 3, 1909, p. 895:2.

78. "Shawmut Car in Race," *Stoneham (Mass.) Independent*, May 29, 1909, n.p.

79. "Bay State Endurance Tie Finally Called a Draw," *Automobile*, Oct. 8, 1908, p. 496:1.

80. "Fast Time Made in Seattle Race," *New York Times*, June 27, 1909, IV, p. 4:6. The last Shawmuts made were 1908 models, according to Beverly Rae Kimes and Henry Austin Clark, Jr., *Standard Catalog of American Cars, 1805–1942*, 2d ed. (Iola, Wis.: Krause

Publications, 1989), p. 1300:1. The fire date is from "Shawmut Plant Succumbs to Flames," *Automobile*, Nov. 19, 1908, p. 712:2.

81. This identification comes from a photo in the "Franklin" photo file at the American Automobile Manufacturers Association in Detroit. The photo shows Gerrie posing in a Franklin car in about 1905.

82. *New York Times*, June 2, 1909, p. 3:5; and "Ocean-to-Ocean Autos Are Here," *Poughkeepsie (N.Y.) Daily Eagle*, June 2, 1909, p. 5:4.

83. *Syracuse (N.Y.) Post-Standard*, June 3, 1909, p. 7:1; and "Race Leaders in Chicago," *Chicago Daily Tribune*, June 6, 1909, III, p. 1:5.

84. "Sixth Car Starts in Seattle Run," *New York Times*, June 5, 1909, p. 7:5.

85. "Seattle Racers Half Way Across," *Automobile Topics*, June 12, 1909, p. 658:2.

86. "Racers to Coast Whiz into Town," *Cleveland Leader*, June 5, 1909, p. 3:2.

87. "Cross-Continent Cars Making Good Time," *Seattle Post-Intelligencer*, June 5, 1909, p. 10:4; *The Reminiscences of C.J. Smith*, p. 9; and *Chronological Story*, p. 12.

88. *Seattle Post-Intelligencer*, June 5, 1909, p. 10:4.

89. Arthur Kelley, "Reading Man Recalls 1909 Auto Race," *Boston Sunday Globe*, June 7, 1959, p. 68:3.

90. *Chicago Daily Tribune*, June 6, 1909, III, p. 1:5; and *Automobile Topics*, June 12, 1909, p. 658:1.

91. Jerry Winters, "Ocean-to-Ocean Auto Contest Is Frenzied, Foolhardy Speed Terror," *Toledo (Ohio) Daily Blade*, June 5, 1909, p. 14:3.

92. "What Light Weight Will Do for You," *Ford Times*, June 15, 1909, p. 5.

93. Untitled blurb, *Ford Times*, July 1, 1909, p. 12:1.

94. "Racers Here To-Night," *Chicago Daily News*, June 5, 1909, p. 2:7 of sports section; and "Automobiles Pass Through," *Goshen (Ind.) Daily News-Times*, June 5, 1909, p. 1:4.

95. *Chicago Daily Tribune*, June 6, 1909, III, p. 1:5; and "With Autoists at Home and Abroad," *South Bend (Ind.) Tribune*, June 7, 1909, p. 9:5.

96. *South Bend (Ind.) Tribune*, June 7, 1909, p. 9:5; and *Chicago Daily Tribune*, June 6, 1909, III, p. 1:5.

97. "Racing Autos Pass the City," *Michigan City (Ind.) Evening News*, June 7, 1909, p. 1:7.

98. "On the Way to Seattle," *Ford Times*, June 15, 1909, p. 3.

99. As identified in "Stearns Car Starts Trip," *South Bend (Ind.) Daily Times*, June 5, 1909, p. 10:4. The 1959 Ford race summary, *Chronological Story of the New York-to-Seattle Transcontinental Endurance Race*, p. 9, without citing sources, names Stolp's crew as Robert Maxwell, driver, Harry W. Sohmers, relief driver, and Kenneth P. Major, mechanic.

100. "New York–Seattle Race Is On," *Motor World*, June 10, 1909, p. 416:3.

101. "The 'Stolp' Shock Eliminator," *Cycle and Automobile Trade Journal*, April 1908, p. 116:1. Another article on Stolp's design appeared as "A New Suspension for Automobiles," *Scientific American*, Jan. 16, 1909, p. 58:2.

102. *New York Times*, June 5, 1909, p. 7:5; *Story of the Race*, p. 6; *Automobile Topics*, June 12, 1909, p. 658:2; *South Bend (Ind.) Daily Times*, June 5, 1909, p. 10:4; and *Chronological Story*, p. 9.

103. *Seattle Post-Intelligencer*, June 2, 1909, p. 1:7.

104. *New York Times*, June 5, 1909, p. 7:5; and "Autos Losing Ground," *Chicago Daily News*, June 7, 1909, p. 6:7.

105. "Auto Racer Down with Diphtheria," *New York Times*, June 7, 1909, p. 8:3.

106. "Route to Seattle Surveyed; All Ready for Event," *Automobile*, May 13, 1909, p. 800:2.

107. *Chicago Daily Tribune*, June 6, 1909, III, p. 1:5, and "Auto Leaders Are Five Hours Ahead," June 7, 1909, p. 13:3.

108. *Chicago Daily News*, June 7, 1909, p. 6:7; and *Chicago Daily Tribune*, June 7, 1909, p. 13:3.

109. *Chicago Daily Tribune*, June 7, 1909, p. 13:3. Throughout this chapter, I'm using the spelling "Whitmore," as did the *Chicago Daily Tribune* and other newspapers. As many more press accounts spelled it "Whittemore," however, and a few said "Whittemon" or even "Whitman."

110. *New York Times*, June 7, 1909, p. 8:3.

111. *New York Times*, June 7, 1909, p. 8:3.

112. *Motor World*, June 10, 1909, p. 416:3; and *Automobile Topics*, June 12, 1909, p. 660:1.

113. The *Transcontinental Automobile Guide*, compiled from Reddington's observations in the Thomas pathfinder, gives the Chicago-St. Louis distance as 317 miles (p. 43).

114. *Chicago Daily News*, June 7, 1909, p. 6:7.

115. "Two Ford Cars, Leaders in National Auto Race, Arrive in Springfield and Continue Journey," *Illinois State Journal* (Springfield), June 7, 1909, p. 2:4.

116. "Racing Autos Pass Through," *Illinois State Register* (Springfield), June 7, 1909, p. 5:5.

117. *New York Times*, June 7, 1909, p. 8:3; and "Racing Automobiles Arrive at St. Louis," *Seattle Post-Intelligencer*, June 8, 1909, p. 8:7.

118. "Car 'Italia' [sic] Goes through City," *Illinois State Register* (Springfield), June 8, 1909, p. 3:7.

119. "The Ocean to Ocean Contest," *Horseless Age*, June 9, 1909, p. 800:2.

120. "Leading Motor Cars Reach Kansas City," *Seattle Post-Intelligencer*, June 9, 1909, p. 6:6. *Automobile Topics*, June 12, 1909, p. 659:2, says the Itala broke not a wheel or steering gear but a steering knuckle.

121. "Seattle Race a Severe Test on Cars," *New York Times*, June 13, 1909, p. 4:7; and "Automobile Law Chart for States Which Regulate," *Automobile*, June 16, 1910, p. 1095.

122. *New York Times*, June 8, 1909, p. 8:5; *Automobile Topics*, June 12, 1909, p. 659:2; and "Ocean to Ocean Racers in the Open West," *Automobile*, June 10, 1909, p. 965:1.

123. "Incidents on Road across Continent," *Seattle Post-Intelligencer*, June 20, 1909, p. 5:1 of sports section. The distance between St. Louis and Glasgow is far less than the 230 miles given in this account.

124. *Automobile*, April 8, 1909, p. 592:2.

125. "Transcontinental Auto Race Trio of 1909 Hold Reunion at Rotary," *Stoneham (Mass.) Independent*, Aug. 17, 1945, n.p.

126. *Seattle Post-Intelligencer*, June 30, 1909, p. 2:5; and "Ford Car Leads the Auto Racers to Denver," *Denver Republican*, June 14, 1909, p. 6:7.

127. From an unidentified Kansas City newspaper clipping reproduced in *Story of the Race*, pp. 16–17.

128. *Story of the Race*, p. 10.

129. Lucy Searfuss, quoted in *The Big Race*, pp. 59-60.

130. *Story of the Race*, p. 10.

131. *Story of the Race*, pp. 11–12.

132. *Chronological Story*, p. 17; and *The Reminiscences of C.J. Smith*, pp. 10–11.

133. "New York–Seattle Race," *Seattle Post-Intelligencer*, June 10, 1909, p. 10:7.

134. "Half Way to Seattle," *New York Times*, June 10, 1909, p. 5:3.

135. *Seattle Post-Intelligencer*, June 10, 1909, p. 10:7.

136. "Acme Car Leads in Race," *New York Times*, June 11, 1909, p. 7:6.

137. "Race Leaders Are Bunched," *Chicago Daily Tribune*, June 12, 1909, p. 9:1.

138. "Ocean to Ocean Cars Encounter Mud and Slush," *Denver Post*, June 12, 1909, p. 1:2.

139. *Denver Post*, June 12, 1909, p. 1:2.

140. "Shawmut Now Has Clear Lead," *Boston Herald*, June 12, 1909, p. 4:5.

141. *Story of the Race*, p. 10.

142. *Denver Post*, June 12, 1909, p. 1:2; and "Seattle Racer in Denver; Mud to Auto's Hubs," *Sunday News-Times* (Denver), June 13, 1909, p. 9:1. (The *Sunday News-Times* is a joint edition of the *Rocky Mountain News* and *Denver Times*.)

143. "Shawmut Car Leads," *Seattle Post-Intelligencer*, June 13, 1909, p. 4:2 of sports section.

144. *Story of the Race*, p. 12. Maps show a "Sandy Creek" near Limon, or about 75 miles from Denver, and—even farther from Denver—a "Sand Creek" between Burlington and Limon.

145. *Sunday News-Times* (Denver), June 13, 1909, p. 9:1.

146. "Mud-Bespattered Ocean to Ocean Racers Arrive," *Denver Post*, June 13, 1909, p. 3:2.

147. *Sunday News-Times* (Denver), June 13, 1909, p. 1:4.

148. "Three Auto Racers Speeding West to Seattle," *Rocky Mountain News* (Denver), June 14, 1909, p. 3:7.

149. *Rocky Mountain News* (Denver), June 14, 1909, p. 3:7. The time separating the two cars is from *Chronological Story*, p. 20.

150. *Denver Republican*, June 14, 1909, p. 6:7.

151. *Story of the Race*, p. 13; and *Denver Republican*, June 14, 1909, p. 6:7.

152. "Coast to Coast Racers Reach Cheyenne," *Wyoming Tribune* (Cheyenne), June 14, 1909, p. 1:1.

153. "Little Fords Sure Winner," *Cheyenne (Wyo.) State Leader*, June 15, 1909, p. 2:3.

154. *Denver Republican*, June 14, 1909, p. 6:7.

155. "Ocean to Ocean Cars Speeding across Wyoming," *Denver Republican*, June 15, 1909, p. 8:3.

156. *Wyoming Tribune* (Cheyenne), June 14, 1909, p. 1:1; and "Vandi Moore, Laramie's Most Enterprising Entrepreneur," *Laramie (Wyo.) Daily Boomerang's* "Annual Edition 1985," July 12, 1985, p. 4:1. Lovejoy dates his first auto in "Lovejoy Takes Hand in Auto Chronology," *Laramie (Wyo.) Republican*, Aug. 15, 1916, p. 6:4.

157. *Wyoming Tribune* (Cheyenne), June 14, 1909, p. 1:1; and *The Big Race*, pp. 71–72.

158. *Cheyenne (Wyo.) State Leader*, June 15, 1909, p. 2:3.

159. *Wyoming Tribune* (Cheyenne), June 14, 1909, p. 1:1.

160. "Ocean to Ocean Cars in Laramie," *Denver Post*, June 14, 1909, p. 5:2.

161. *Story of the Race*, p. 13.

162. "Ford Car at Rawlins, Wyo.," *New York Times*, June 15, 1909, p. 9:6.

163. "Ford Takes Lead across Continent," *Seattle Post-Intelligencer*, June 15, 1909, p. 8:4.

164. "Transcontinental Racers in Denver," *Denver Times*, June 14, 1909, p. 3:2.

165. The Shawmut's Monday arrival and Tuesday departure times are from "Shawmut Car Leads Racers," *New York Times*, June 16, 1909, p. 9:6. Other accounts offer slightly different times. The history of Fort Steele is from *Wyoming: A Guide to its History, Highways, and People* (New York: Oxford University Press, 1941), pp. 238–39.

166. *Motor World*, June 17, 1909, p. 457:3.

167. *New York Times*, June 15, 1909, p. 9:6; and *Story of the Race*, p. 13.

168. *New York Times*, June 16, 1909, p. 9:6.

169. "Itala Car Hit by a Train," *Chicago Daily Tribune*, June 15, 1909, p. 9:1.

170. *Chronological Story*, pp. 18–19.

171. *Story of the Race*, p. 14.

172. "Shawmut Leading Field," *Boston Herald*, June 16, 1909, p. 5:8; *Chronological Story*, p. 22; and *New York Times*, June 16, 1909, p. 9:6.

173. *Boston Herald*, June 16, 1909, p. 5:8; and "Ocean to Ocean Race," *Motor Age*, June 17, 1909, p. 12:3.

174. "Ford Car in Lead," *Montpelier (Idaho) Examiner*, June 18, 1909, p. 2:1.

175. "Three Seattle Racing Cars Bunched," *New York Times*, June 17, 1909, p. 8:7.

176. *Story of the Race*, p. 15; and "Both Ford Cars Speeding Ahead," *Seattle Post-Intelligencer*, June 18, 1909, p. 10:5.

177. *Story of the Race*, p. 15; and *Chronological Story*, p. 23.

178. *Montpelier (Idaho) Examiner*, June 18, 1909, p. 2:1; and "Racers Speeding In Idaho," *New York Times*, June 18, 1909, p. 5:5.

179. "Ocean-to-Ocean Racers Arrive in the Gate City," *Pocatello (Idaho) Tribune*, June 18, 1909, p. 1:6.

180. *Story of the Race*, p. 15.

181. *Pocatello (Idaho) Tribune*, June 18, 1909, p. 1:6. The Oregon Short Line Railroad ran from Granger, Wyo., to Portland, Ore.

182. *Pocatello (Idaho) Tribune*, June 18, 1909, p. 1:6.

183. "Acme Auto Finds Cheyenne; Ford No. 1 at Pocatello," *Wyoming Tribune* (Cheyenne), June 18, 1909, p. 1:6.

184. *Story of the Race*, p. 18.

185. "Those Auto Racers," *Pocatello (Idaho) Tribune*, June 19, 1909, p. 1:3.

186. "Wise and Otherwise," *Shoshone (Idaho) Journal*, June 25, 1909, p. 7:4.

187. *Chronological Story*, p. 24.

188. "The Winning of the Race," *Ford Times*, July 15, 1909, p. 5.

189. *The Reminiscences of C.J. Smith*, p. 10.

190. *Pocatello (Idaho) Tribune*, June 19, 1909, p. 1:3.

191. "First Car in Race Reaches Boise," *Idaho Daily Statesman* (Boise), June 20, 1909, p. 2:3.

192. "Ford Cars Keep to Fore," *Boston Herald*, June 19, 1909, p. 6:6.

193. *Idaho Daily Statesman* (Boise), June 20, 1909, p. 2:3.

194. *The Reminiscences of C.J. Smith*, p. 11.

195. "Neck and Neck from New York," *Seattle Post-Intelligencer*, June 20, 1909, p. 5:5 of sports section.

196. "Racers Pass Boise," *Idaho Daily Statesman* (Boise), June 21, 1909, p. 3:2.

197. *Idaho Daily Statesman* (Boise), June 21, 1909, p. 3:2.

198. "Ford No. 2 in the Lead," *New York Times*, June 21, 1909, p. 8:4; and "Ocean-to-Ocean Automobiles Have Registered at Baker City," *Baker City (Ore.) Herald*, June 21, 1909, p. 1:3.

199. *Baker City (Ore.) Herald*, June 21, 1909, p. 1:3.

200. "Ford No. 2 Winner of Great Auto Race," *Seattle Post-Intelligencer*, June 24, 1909, p. 10:6.

201. "The Ocean to Ocean Contest," *Horseless Age*, June 23, 1909, p. 860:3.

202. "Ford No. 2 Leading Near End of Race," *New York Times*, June 22, 1909, p. 8:5.

203. "Ford No. 2 Leads in Hot Auto Race," *Seattle Post-Intelligencer*, June 22, 1909, p. 10:7.

204. "Shawmut Gaining," *La Grande (Ore.) Evening Observer*, June 22, 1909, p. 1:3.

205. *New York Times*, June 22, 1909, p. 8:5.

206. *Story of the Race*, pp. 19-20.

207. *Seattle Post-Intelligencer*, June 22, 1909, p. 10:6.

208. "Ford Car Has Fire," *Baker City (Ore.) Morning Democrat*, June 22, 1909, p. 1:3; *Story of the Race*, p. 21; and "Ford Victor Fire-Scarred," *Automobile*, July 29, 1909, p. 203:1.

209. "Ford Is Leading," *Baker City (Ore.) Herald*, June 22, 1909, p. 8:4.

210. "Builder of the Fords Goes Out to Meet Cars," *Seattle Post-Intelligencer*, June 23, 1909, p. 1:5.

211. *Seattle Post-Intelligencer*, June 23, 1909, p. 1:5.

212. *Seattle Post-Intelligencer*, June 23, 1909, p. 1:5.

213. "Result of Race Is in Doubt," *Idaho Daily Statesman* (Boise), June 23, 1909, p. 2:3.

214. "Three New York Cars Near Seattle," *New York Times*, June 23, 1909, p. 9:1.

215. *Seattle Post-Intelligencer*, June 24, 1909, p. 10:5.

216. *The Reminiscences of C.J. Smith*, pp. 11-12.

217. "Ford Car Wins Race," *Pocatello (Idaho) Tribune*, June 23, 1909, p. 1:2; "Ford No. 2 Wins Race," *La Grande (Ore.) Evening Observer*, June 23, 1909, p. 1:6; and "Ford No. 2 Wins Great Race," *Baker City (Ore.) Herald*, June 23, 1909, p. 1:1.

218. "Ford Car First Machine to Complete Trip to Seattle," *Chicago Daily Tribune*, June 24, 1909, p. 13:6.

219. *Seattle Post-Intelligencer*, June 24, 1909, p. 10:6.

220. *The Reminiscences of C.J. Smith*, p. 12.

221. *Seattle Post-Intelligencer*, June 24, 1909, p. 10:5.

222. *New York Times*, June 27, 1909, IV, p. 4:6.

223. *Ford Times*, July 15, 1909, p. 1.

224. The distance—higher than the Thomas Flyer's 4,001 miles to Seattle—is from *Story of the Race*, p. 19.

225. "Ford Car Crosses Continent with Air from New York Still in Tires," *Seattle Post-Intelligencer*, June 27, 1909, p. 2:3 of real estate section. Exactly how many tires Ford No. 2 used is an open question. Henry Ford claimed the Ford racers would carry no spares. In *The Big Race*, pp. 61 and 83, Jack B. Scott says the Fords "were going to restock with tires and oil in Kansas City," and later quotes a spectator who recalls that Ford No. 2 came through Sunnyside, Wash., carrying with it "a dozen tires from ragged to new"—an improbably large number. The *Illinois State Register* (Springfield), June 7, 1909, p. 5:5, said the Fords had "tires and rims along in case of emergency."

226. *The Reminiscences of C.J. Smith*, pp. 13–14. Babbitt, a relatively soft alloy of antimony, tin, lead and copper, is most commonly used for engine crankshaft and connecting-rod bearings, which are subject to regular, consistent forces. Smith's point is that the uneven jolting and jarring of rough roads tended to flatten or elongate such a soft bearing, which is why the automaker switched to a roller bearing that ran in a hardened-steel race.

227. *Motor World*, June 24, 1909, p. 500:3.

228. *Seattle Post-Intelligencer*, June 24, 1909, p. 10:5.

229. *Seattle Post-Intelligencer*, June 24, 1909, p. 10:6.

230. *Seattle Post-Intelligencer*, June 24, 1909, p. 10:6; and *Boston Sunday Globe*, June 7, 1959, p. 68:8.

231. *Chicago Daily Tribune*, June 24, 1909, p. 13:6.

232. "Shawmut Car on A.Y.P. Grounds," *La Grande (Ore.) Evening Observer*, June 24, 1909, p. 1:5.

233. *Story of the Race*, p. 22.

234. *Chronological Story*, p. 29.

235. "Protest Heard in Automobile Race," *Seattle Post-Intelligencer*, June 26, 1909, II, p. 4:5.

236. "Ford Car No. 2 Gets Prize," *New York Times*, June 30, 1909, p. 9:2.

237. "Ford Cars Are Under a Cloud," *Baker City (Ore.) Herald*, June 26, 1909, p. 1:3.

238. *Pocatello (Idaho) Tribune*, June 24, 1909, p. 1:1.

239. *New York Times*, June 26, 1909, p. 8:7.

240. Information and quotations about the charges come from the *Seattle Post-Intelligencer*, "Shawmut Car Second in Race Enters Protest," June 25, 1909, p. 1:3, and June 26, 1909, II, p. 4:5; and "Protest of Ford Car No. 2," *New York Times*, June 26, 1909, p. 8:7.

241. *Seattle Post-Intelligencer*, June 26, 1909, II, p. 4:5.

242. *Seattle Post-Intelligencer*, June 26, 1909, II, p. 4:5.

243. *Seattle Post-Intelligencer*, June 26, 1909, II, p. 4:5.

244. *Seattle Post-Intelligencer*, June 26, 1909, II, p. 4:5.

245. *Seattle Post-Intelligencer*, June 26, 1909, II, p. 4:5.

246. *Ford Times*, July 15, 1909, p. 7.

247. *Seattle Post-Intelligencer*, June 30, 1909, pp. 1:3, 2:2.

248. "Autos and Autoists," *Detroit Saturday Night*, Nov. 27, 1909, p. 5:2.

249. "Coast to Coast Race Shawmut's," *Motor Age*, Nov. 4, 1909, p. 55:1.

250. "Big Automobile Trophy Awarded to Shawmut Car," *Seattle Post-Intelligencer*, Oct. 30, 1909, p. 10:1.

251. "Dispute Over Long Race," *Automobile*, April 21, 1910, p. 767:1.

252. *The Reminiscences of C.J. Smith*, p. 12.

253. Philip Van Doren Stern, *Tin Lizzie: The Story of the Fabulous Model T Ford* (New York: Simon and Schuster, 1955), p. 74.

254. From an unidentified Kansas City newspaper clipping reproduced in *Story of the Race*, pp. 16–17.

255. "Motors and Motorists," *Kansas City (Mo.) Star*, Sept. 5, 1909, p. 11A:3. Background information on the return trip is from the same *Star* column, Aug. 22, 1909, p. 10A:2; and from *Story of the Race*, p. 23.

256. Ford press release in "Ford Caravan Marking Famous Race Will Arrive on Saturday," *Wyoming State Tribune* (Cheyenne), June 11, 1959, p. 4:1; and "Ford Re-Enacts 1909 Endurance Contest," *Northern Automotive Journal*, July 1959, p. 24:2.

257. *Northern Automotive Journal*, July 1959, p. 24:1.

258. David L. Lewis, "Ford and the Road of the Racer—Part II," *Ford Life*, July-August 1971, p. 25:3.

259. This the author discovered while searching for photos and information during a January 1996 visit to the research center.

260. *Stoneham (Mass.) Independent*, Aug. 17, 1945, n.p.

261. "Robert H. Messer Conducted Local Funeral Service for Many Years," *Stoneham (Mass.) Independent*, Oct. 9, 1952, p. 1.

262. *Stoneham (Mass.) Independent*, "Sales to be Ready First of the Year," Nov. 15, 1946, n.p., and "T.A. Pettengill Was Pioneer in Auto Industry," March 10, 1955, n.p.

263. "Earle H. Chapin," *Reading (Pa.) Chronicle*, March 19, 1964, p. 4.

264. John Gunnell, "Model 'T' Club's Dream Run Departs from New York City," *Old Cars Weekly*, Aug. 9, 1984, p. 1.

265. Joseph C. Ingraham, "Vintage Cars Wheeze Off Here to Re-enact 23-Day Drive of '09," *New York Times*, June 2, 1959, p. 40:3.

266. "Disgraced by Our Bad Roads," editorial, *New York Times*, Feb. 17, 1908, p. 6:4.

267. See Emily Post, *By Motor to the Golden Gate* (New York: D. Appleton and Co., 1916), p. 67.

268. Henry B. Joy, "Transcontinental Trails: Their Development and What They Mean to This Country," *Scribner's Magazine*, February 1914, pp. 160, 161.

269. "Huss, Lionized at Portland, Talks of His Long Race," *Motor World*, June 29, 1905, p. 607:3.

270. "One Man's Family Tour from Coast to Coast," *Automobile*, May 28, 1908, p. 733:2.

271. "Long Ocean to Ocean Trail Is Blazed," *Motor Age*, Aug. 17, 1911, p. 8:3; and "Ocean-to-Ocean Tour's Lessons," *Automobile Topics*, Dec. 9, 1911, p. 191:1.

272. From a 25-page unpublished typewritten history of Dwight Huss' experiences, *Chronology of [the] 1905 Olds Transcontinental Race*, p. 7, shared by Dwight's son, John Huss.

273. "Impressions of a Pilot," *New York Times*, March 8, 1908, II, p. 2:5; and "Big Auto Race Puts Life into Auto Race," *Omaha (Neb.) World-Herald*, March 16, 1908, p. 2:3.

274. Oldsmobile ad, *MoTor*, July 1905, p. 15:1.

275. "Roberts May Enter New Race," *New York Times*, May 20, 1908, p. 1:1.

Appendix:

Transcontinental Races at a Glance

1905 OLDSMOBILE RACE

Drivers	Dwight B. Huss and Milford Wigle
Car	1-cylinder, 7-horsepower, 1904 Oldsmobile runabout, "Old Scout"
City-City	New York City–Portland, Ore.
Dates	9:30 a.m. May 8–12:58 p.m. June 21, 1905
Distance	3,500 to 4,000 miles
Elapsed Time (ET)	44 days, 6 hours, 28 minutes
Running Time	42 out of 45 calendar days
Average Speed (ET)	3.53 mph (for 3,750 miles)
Firsts	One of two cars in the first transcontinental race and the first east-to-west crossing. Oldsmobile becomes the first make of auto to cross the continent a second and third time, the first being in 1903.

Drivers	Percy F. Megargel and Barton Stanchfield
Car	1-cylinder, 7-horsepower, 1904 Oldsmobile runabout, "Old Steady"
City-City	New York City–Portland, Ore.
Dates	9:30 a.m. May 8–5 p.m. June 28, 1905
Distance	3,500 to 4,000 miles
Elapsed Time (ET)	51 days, 10 hours, 30 minutes
Running Time	48 out of 52 calendar days
Average Speed (ET)	3.04 mph (for 3,750 miles)
Firsts	One of two cars in the first transcontinental race and the first east-to-west crossing. Oldsmobile becomes the first make of auto to cross the continent a second and third time, the first being in 1903.

1908 NEW YORK–PARIS RACE (IN AMERICA)

Thomas Auto (U.S.)

Drivers	Harold S. Brinker, E. Linn Mathewson, Montague "Monty" Roberts, George Schuster

Crew	Charlie Duprez, Hans Hendrik Hansen*, Mason B. Hatch, George Miller, T. Walter Williams
Car	4-cylinder, 70-horsepower, 1907 Model DX Thomas Flyer roadster
City-City	New York City–San Francisco
Dates	11:15 a.m. Feb. 12–4:40 p.m. March 24, 1908
Distance	3,836 miles
Elapsed Time (ET)	41 days, 8 hours, 25 minutes
Average Speed (ET)	3.87 mph
Firsts	New York–Paris entrant that won the first U.S. transcontinental race involving more than one make of auto.

---

*Hansen traveled part of the way across America in the De Dion racer.

Zust Auto (Italian)

Drivers	Henri Haaga, Emilio Sirtori
Crew	A.L. Ruland, Antonio Scarfoglio
Car	4-cylinder, 30- to 40-horsepower Zust
City-City	New York City–San Francisco
Dates	11:15 a.m. Feb. 12–10:35 a.m. April 4, 1908
Distance	about 4,600 miles
Elapsed Time (ET)	52 days, 2 hours, 20 minutes
Average Speed (ET)	3.68 mph
Firsts	New York-Paris entrant that finished second in the first U.S. transcontinental race involving more than one make of auto.

De Dion Auto (French)

Drivers	Alphonse Autran, Emanuel Lescares, G. Bourcier St. Chaffray
Crew	Hans Hendrik Hansen*
Car	4-cylinder, 30- to 40-horsepower De Dion
City-City	New York City–San Francisco
Dates	11:15 a.m. Feb. 12–5:30 p.m. April 7, 1908
Distance	about 4,090 miles
Elapsed Time (ET)	55 days, 9 hours, 15 minutes
Average Speed (ET)	3.08 mph
Firsts	New York–Paris entrant that finished third in first U.S. transcontinental race involving more than one make of auto.

---

*Hansen traveled part of the way across America in the Thomas racer.

1909 NEW YORK–SEATTLE RACE

Cars and Drivers   **Acme (6 cyl, 48 hp)**: James A. Hemstreet, Jerry Price, George Salzman, Fay R. Sheets
Ford No. 1 (4 cyl, 20 hp): H.B. Harper, Frank Kulick
Ford No. 2 (4 cyl, 20 hp): B.W. (Bert) Scott, C.J. (Jimmy) Smith
Itala (4 cyl, 45 hp): Elbert Bellows, Gus Lechleitner (New York City–Chicago), Egbert Lillie (St. Louis-Cheyenne), F.B. Whitmore
Shawmut (4 cyl, 40 hp): E.H. (Earle) Chapin, R.H. (Robert) Messer, T. Arthur Pettengill

City-City   New York City–Seattle

Dates, Elapsed Time (ET)   <u>Start</u>: 3 p.m. June 1, 1909

<u>Finish</u>:

*Ford No. 2	12:55 p.m.	6/23/09	(22d-00h-55m)
Shawmut	5:33 a.m.	6/24/09	(22d-17h-33m)
*Ford No. 1	11:58 a.m.	6/25/09	(23d-23h-30m)
Acme	3:24 p.m.	6/29/09	(28d-03h-24m)
Itala	—withdrawn—		

Distance   4,106 miles

Average Speed (ET)

Ford No. 2	7.76 mph
Shawmut	7.53 mph
Ford No. 1	7.13 mph
Acme	6.08 mph

Firsts   The fastest transcontinental race to that time.

---

*Both Fords disqualified. Shawmut awarded 1st place, Acme 2nd.

Index

About the Author

CURT McCONNELL has written about automotive history for *Automobile Quarterly,* the Horseless Carriage Club of America's *Horseless Carriage Gazette* and the Society of Automotive Historians' *SAH Journal.* His articles on automotive history and other subjects have appeared in the *Chicago Tribune, Washington Post, Cleveland Plain Dealer, Denver Post, Christian Science Monitor* and other major newspapers, as well as in *The*

Lincoln Highway Forum (Lincoln Highway Association), *Wheels* (Detroit Public Library's National Automotive History Collection) and *Nebraska History* (Nebraska State Historical Society). His writing has also appeared in the magazines of the Plymouth Owners Club and the National DeSoto Club. In addition to his work for newspapers and magazines, McConnell wrote the automotive section of the *Encyclopedia of the Great Plains,* published by the Center for Great Plains Studies at the University of Nebraska-Lincoln.

McConnell's first book, *Great Cars of the Great Plains,* a study of the contributions made by five Midwestern manufacturers in the early days of the auto industry, won the Antique Automobile Club's 1995 Thomas McKean Memorial Cup for historical research.

To learn firsthand the rigors of transcontinental travel, McConnell packed his 1939 Plymouth coupe and set off to drive across America on dirt and gravel roads. He is writing about his discoveries for an upcoming book.

In addition to Midwestern autos, transcontinental auto trips and Chrysler products, McConnell is interested in early highways and engine makers. He lives near Lincoln, Nebraska.